ANCIENT SEA REPTILES

Plesiosaurs, ichthyosaurs, mosasaurs and more

Darren Naish

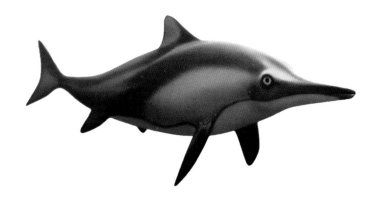

Smithsonian Books
Washington, DC

First published by the Natural History Museum, Cromwell Road, London SW7 5BD © The Trustees of the Natural History Museum, London, 2022

The Author has asserted their right to be identified as the Author of this work under the Copyright, Designs and Patents Act 1988

All rights reserved. No part of this publication may be transmitted in any form or by any means without prior permission from the publishers.

Published in North America, South America, Central America, and the Caribbean by Smithsonian Books

This book may be purchased for educational, business, or sales promotional use. For information, please write: Special Markets Department, Smithsonian Books, P.O. Box 37012, MRC 513, Washington, DC 20013

Library of Congress Cataloging-in-Publication Data
Names: Naish, Darren, author.
Title: Ancient sea reptiles : plesiosaurs, ichthyosaurs, mosasaurs, and more / Darren Naish.
Description: Washington, DC : Smithsonian Books, [2023] | Includes bibliographical references and index.
Identifiers: LCCN 2022032265 | ISBN 9781588347275 (hardcover)
Subjects: LCSH: Marine reptiles, Fossil. | Paleontology--Mesozoic.
Classification: LCC QE861 .N35 2023 | DDC 567.9/37--dc23/eng/20220706
LC record available at https://lccn.loc.gov/2022032265

Printed in China, not at government expense
27 26 25 24 23   2 3 4 5

Designed by Bobby Birchall, Bobby&Co.
Reproduction by Saxon Digital Services
Printed by 1010 Printing International Limited, China

CONTENTS

1	Introduction	4
2	Evolution	28
3	Anatomy	36
4	The lesser-known groups: mesosaurs, Triassic sauropterygians, Cretaceous sea snakes and more	54
5	Shark-shaped reptiles: the ichthyosaurs and their kin	94
6	Long necks, big mouths: the plesiosaurs	124
7	Sea crocs: the thalattosuchians	152
8	Mosasaurs: the great sea lizards	164
9	Sea turtles	178

Glossary	186
Further information	187
Index	188
Picture credits and acknowledgements	192

1 | INTRODUCTION

BETWEEN ABOUT 252 AND 66 million years ago the seas, oceans and estuaries of the world were home to an amazing array of giant reptiles. It was the geological time known as the Mesozoic, or Age of Reptiles, which consisted of the Triassic, Jurassic and Cretaceous. These giant reptiles, known collectively as Mesozoic marine reptiles, occupied the roles filled today by seals, dolphins and whales and resembled them in aspects of appearance, behaviour and biology.

There were five main Mesozoic marine reptile groups: the shark-shaped ichthyosaurs, the (mostly) long-necked plesiosaurs, the crocodile-like thalattosuchians or sea crocs, the giant swimming lizards known as mosasaurs, and the sea turtles. All five groups have complicated evolutionary histories and did not all exist at the same time. As we'll see, a diversity of lesser, shorter-lived groups existed alongside the main cast.

Mesozoic marine reptiles ruled the seas at about the same time as dinosaurs were the dominant animals on land. They were not dinosaurs, however, nor closely related to them. It remains true, however, that Mesozoic marine reptiles are often considered 'default dinosaurs' by the public and an argument can be made that part of our love affair with dinosaurs is due to Mesozoic marine reptiles – plesiosaurs in particular – being regarded, albeit incorrectly, as part of the group.

In general, Mesozoic marine reptiles were streamlined, equipped with paddle- or wing-shaped limbs, and often with a tail built to produce thrust in

water. They were similar in size to marine mammals, the smallest being around 1 m (3¼ ft), the biggest exceeding 20 m (66 ft). Most were predators, with jaws and teeth built to grab, pierce and cut, but some were specialized for a diet of shellfish, for filtering small animals from mud or water, or for extracting prey from burrows or crevices. Bite marks on the bones of other marine reptiles, stomach contents, injured jaws and faces and even marks left in solidified mud provide clues to Mesozoic marine reptile behaviour and lifestyle.

Fossils of these animals have been found across the globe, which is not surprising given that they swam throughout the seas of the world. Early discoveries in Europe and North America were pivotal in helping people understand the history of life on Earth. They demonstrated that animals existed in the past that are unlike those of today, and provided evidence for continental movement. They showed that ancient seas and sea levels were different from those of today, and even shed light on the concepts of extinction and evolution. The discovery of Mesozoic marine reptiles was key to the recognition of a sea that cut North America in two during the Cretaceous as well as the existence of another that flooded Europe during the Mesozoic, the Tethys Ocean.

Mesozoic marine reptiles have been so modified by evolution that experts have struggled to work out where they fit in the reptile family tree. We know they represent at least four separate invasions of the water: thalattosuchians

> Over the nearly 190 million years of the Mesozoic Era (and during the later part of the Paleozoic too), reptiles evolved an impressive diversity of marine lineages. These varied substantially in shape, and in size too. Most were between 1 and 7 m (3–23 ft) in length but giants exceeding 10 m (33 ft) evolved on many occasions.

The Victorian marine reptile models (opposite) of Crystal Palace, London, show how scientists imagined these animals to look during the 1850s. These photos show (clockwise from top) a long-necked plesiosaur, a teleosaurid thalattosuchian, and the mosasaur *Mosasaurus*.

Ichthyosaurs like this Jurassic *Ophthalmosaurus* had slender jaws and enormous eyes. When alive, muscle, blubber and skin made it a superbly streamlined animal. This skeleton lacks teeth but this is because they've fallen out of the jaws – the live animal possessed them.

are part of the crocodylomorph group, mosasaurs are lizards, ichthyosaurs and plesiosaurs – while of less certain affinity – represent an entirely separate group, and sea turtles are obviously part of the turtle lineage.

The chapters that follow will tackle the story of these ancient beasts, covering the new theories, discoveries and points of controversy around this fascinating subject.

INTRODUCING THE MAIN GROUPS

The term Mesozoic marine reptile applies to a diversity of animals that vary substantially in shape and behaviour. Unfortunately, most of them lack common names, meaning that there is no way to discuss them without using their technical and sometimes cumbersome scientific ones. One of the first things to do in this book, then, is introduce the main characters of the story.

Firstly, there are the ichthyosaurs, sometimes called fish lizards or fish reptiles and noted for their dolphin- or shark-like shape. A typical ichthyosaur was a streamlined reptile, 2 or 3 m (6½ or 10 ft) long with slender jaws, conical teeth, huge eyes, two pairs of flippers, a dorsal fin, and a tail with a vertical, crescent-shaped fin. We know much about how they looked thanks to specimens that preserve their soft-tissue outline. Not all ichthyosaurs were this shape, though. In fact, dolphin- and shark-shaped ichthyosaurs (known as parvipelvians) only evolved at the end of the Triassic. Older kinds looked different, variously having shorter jaws, longer bodies, tails that lacked a crescent-shaped fin, and flattened or dome-shaped teeth.

Secondly, there are the plesiosaurs, a group as popular as the dinosaurs. Plesiosaurs are famous for their long necks and small heads, but several

groups – including the pliosaurids – include species with short-necks and enormous heads recalling those of crocodiles. Numerous species sit between these extremes, too. All plesiosaurs share the same anatomical plan: the body is relatively stiff, the tail is short, and propulsion was provided by four wing-like flippers. Plesiosaurs are closely related to several Triassic groups: the armoured placodonts, the small pachypleurosaurs, the mostly long-jawed nothosaurs, and the plesiosaur-like pistosaurs. Pachypleurosaurs and some nothosaurs were likely capable of movement on land. These groups were mostly, but not entirely, restricted to the Tethys Ocean, the sea that stretched from Europe to China, and most were 1 to 4 m (3¼ to 13 ft) long. All are united with plesiosaurs within Sauropterygia, though plesiosaurs are the only sauropterygians that persisted beyond the Triassic.

Next are the thalattosuchians, sometimes called sea crocs, which is what thalattosuchian means. Thalattosuchians belong to Archosauria, the ruling reptile group that includes dinosaurs, and specifically within Crocodylomorpha, the group that includes crocodylians. Thalattosuchians originated close to the Triassic-Jurassic boundary (around 200 million years ago) and included armour-plated, long-snouted forms that would have looked like crocodiles and gharials as well as the more aquatic metriorhynchids. These lacked scales and possessed paddle-like limbs and a vertical tail fin. They were important throughout the Jurassic but died out by the end of the Early Cretaceous, 120 million years ago.

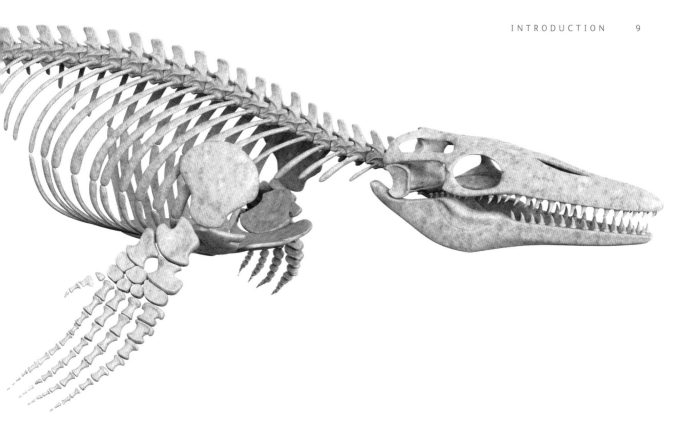

Then, come the mosasaurs. Mosasaurs are squamates: part of the lizard and snake group, Squamata. They are also the last Mesozoic marine reptile group to evolve and (relatively speaking) the shortest-lived, their time of existence being limited to the last 40 million years of the Cretaceous. Early mosasaurs were 1 m (3¼ ft) long or less and probably capable of foraging on land. But by 90 million years ago, they had evolved into giants, 5 m (16 ft) and more, with flipper-shaped limbs and a tail fin, and fully aquatic.

Lastly, there are the sea turtles, a group that still exists but includes a number of spectacularly big and odd Mesozoic species that lived alongside plesiosaurs and mosasaurs.

These groups were the longest-lived in geological terms, and the most spectacular to feature in this story. But a number of other, more obscure, groups also need coverage, the ones that tend to be forgotten in films or museum displays. They include the Triassic hupehsuchians and thalattosaurs, the turtle-like saurosphargids, and the long-necked tanystropheids.

The mosasaur skeleton – this belongs to the giant Late Cretaceous form *Mosasaurus* – has the 'open' rear skull structure typical of lizards. A long, flexible body and tail was typical too, as was a size substantially exceeding that of terrestrial lizards. The downturned end part of the tail supported a vertical fluke.

THE WORLD OF MESOZOIC MARINE REPTILES

The 186 million years of the Mesozoic was such a long stretch of time that making generalizations about it is difficult. The shapes and arrangements of the continents changed, as did the locations and sizes of seas and oceans. Earth during the Mesozoic was in a 'hothouse' phase. This was partly due

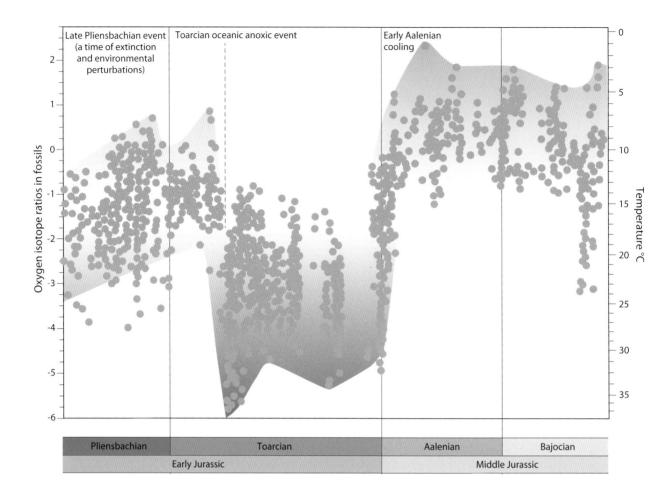

The view that Jurassic temperatures were constantly high is challenged by geochemical evidence. This graph shows data obtained from Jurassic shelly fossils found across Europe. They show that seawater temperatures were sometimes extremely high, but that they were cool and even cold during parts of the Early and Middle Jurassic.

to the unified form of the continents, which led to dry continental interiors and small coastlines. There was no land over the poles, which normally reflects sunlight and thus cools the planet, and atmospheric CO_2 was high, trapping heat.

Hothouse worlds have little ice, and a consequence of little ice is high sea levels. Mesozoic sea levels were the highest in history, and low-lying areas were flooded. Vast continental and continental shelf regions were covered by warm, shallow seas, few more than 60 m (197 ft) deep, with surface temperatures that could exceed 30°C (86°F) – today's average is 18°C (64.5°F). The seafloor in these shallow areas was within the reach of sunlight, so algae could grow and complex communities could thrive. The oceans were fertile places, as widespread volcanic activity and wildfires on land increased chemical runoff of iron, nitrogen and other chemical elements into the seas.

The combination of warmth, shallowness and chemical richness made the Mesozoic seas the ideal environment for marine reptiles to evolve. Indeed, these animals exploded in diversity from the Triassic, at the start of the

Mesozoic, onwards. The high temperatures and abundance of resources encouraged rapid growth and rapid evolution, while the complexity and richness of the different marine habitats encouraged specialization and diversification.

However, times were not always good. On several occasions during the Mesozoic, volcanic eruptions released huge quantities of carbon into the oceans, turning them acidic, and reducing their oxygen content. This killed large swathes of micro-organisms, causing food chains to collapse. Mass extinctions then followed.

The shape of Mesozoic seas

During the Triassic, between 251 and 201 million years ago, the continents were united as the supercontinent Pangaea, and just two major bodies of water existed. The Tethys Ocean was partially enclosed to its north and east by areas that today form China, and it also extended from eastern North America and northern Africa all the way to Australasia. The rest of the globe, around 70% of it, was covered by the ocean Panthalassa. Open ocean was thus a major feature of the Triassic world.

During the Jurassic, between 201 and 145 million years ago, continental fragmentation and high sea levels – sometimes more than 100 m (328 ft) higher than those of today – resulted in the creation of extensive shallow seas. Most of Europe was flooded, only the higher areas remaining emergent as islands. Continental shallows surrounded the north, east and south of the southern continents, all of which were united as Gondwana. On North America's west coast, a tectonically active zone resulted in chains of volcanic islands, marine trenches and sea volcanoes. This created fertile, complex

Eon	Era	Period	Epoch	Age	
Phanerozoic	Mesozoic	Cretaceous	Upper	Maastrichtian	66
				Campanian	72.1
				Santonian	83.8
				Coniacian	86.3
				Turonian	89.8
				Cenomanian	93.9
					100.5
			Lower	Albian	113.0
				Aptian	126.3
				Barremian	130.8
				Hauterivian	133.9
				Valanginian	139.4
				Berriasian	145.0
		Jurassic	Upper	Tithonian	152.1
				Kimmeridgian	157.3
				Oxfordian	163.5
			Middle	Callovian	166.1
				Bathonian	168.3
				Bajocian	170.3
				Aalenian	174.1
			Lower	Toarcian	182.7
				Pliensbachian	190.8
				Sinemurian	199.3
				Hettangian	201.3
		Triassic	Upper	Rhaetian	208.5
				Norian	229.4
				Carnian	237.0
			Middle	Ladinian	241.5
				Anisian	247.1
			Lower	Olenekian	250.0
				Induan	

Geological time is split up into a tiered system of subdivisions. The animals in this book lived during that part of the Phanerozoic (the age of 'visible life') termed the Mesozoic, and individual species were mostly restricted to specific Ages. The numbers refer to millions of years ago.

The supercontinent Pangaea stretched from north to south during the Triassic and the early part of the Jurassic. The formation of rift valleys between North America and Africa caused Pangaea to tear in two during the middle of the Jurassic, around 175 million years ago.

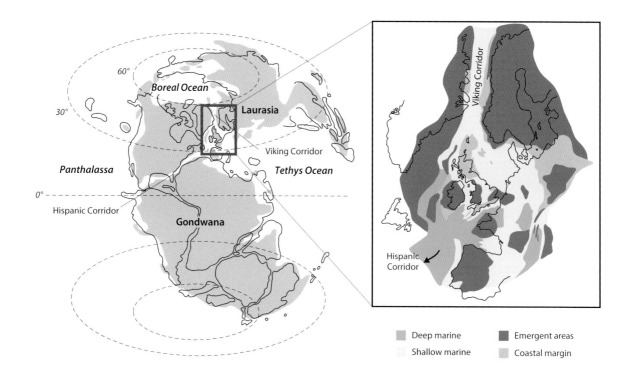

Deep marine | Emergent areas
Shallow marine | Coastal margin

> The semi-enclosed Boreal Ocean was a feature of the far north during the Jurassic. It had high-latitude connections both to Panthalassa and Tethys, the latter via the Viking Corridor. Some fossil evidence suggests that Boreal Ocean ecosystems were specialized and unique.

environments, which persisted right to the end of the Cretaceous. An inlet of Panthalassa formed an intercontinental sea – the Sundance Sea – which flooded the land from British Columbia to Wyoming almost 2,000 km (1,243 m) inland. Ichthyosaurs and plesiosaurs flourished here.

As Gondwana moved south and the northern continents (united as Laurasia) moved north, a seaway emerged, connecting Europe with eastern Panthalassa. This is the Hispanic Corridor or Caribbean Seaway and it probably explains why Europe, Mexico and the Caribbean shared similar Jurassic marine reptiles. Another seaway – the Mozambique Corridor or Trans-Erythraean Seaway – allowed Tethyan animals to move southwest and into a region between Antarctica and the conjoined Africa and South America. Metriorhynchids, parvipelvians and plesiosaurs related to European forms existed in western Argentina during the Late Jurassic. It's possible they used the Hispanic or Mozambique corridors to get there.

Meanwhile, the Viking Corridor connected the Western Tethys Ocean with a semi-enclosed sea in the north, the Boreal Ocean, also called the Boreal Sea or Arctic Ocean. Despite its Arctic Circle location, which caused low temperatures and seasonal darkness, ichthyosaurs and plesiosaurs made it their home.

Our knowledge of Jurassic marine reptiles is dominated by information from Europe, and it's mostly for this reason that we've generally imagined Europe as the area where these groups originated. This is a consequence of human history and the use of Jurassic rocks in industry, but it's also because European Jurassic sediments are on the land surface today and have survived erosion and subduction. The complexity of the Jurassic world demonstrates that other regions were important in Mesozoic marine reptile history and may have been the true origination sites of these groups, it's just that the fossil record of those places is not as well understood or as well sampled.

As the Jurassic ended and the Cretaceous began, Gondwana fragmented even further, Antarctica moved south, and North America's shape changed as a new continental sea formed. India became an island and moved north, but modern India is only part of 'Greater India', the northern edge of which was more than 2,500 km (1,553 miles) north of the region's modern margin. This was an enormous submerged continental area. Further east, another submerged continental region – Zealandia or Greater New Zealand – existed in the southwest Pacific. Some of the animals known as fossils here occurred around Antarctica as well as southern and western South America. The ecosystems across this area formed a biogeographic region known as the Weddellian Province. Weddellian animals must have been cold tolerant since they lived in seas where ice formed during winter.

The opening of the South Atlantic between South America and Africa was a key event in the Cretaceous. This produced more coastline and provided another route for north-south movement of sea creatures. By 75 million years ago, mosasaurs known from Europe and North America lived around Angola on Africa's west coast, as well as around Brazil on the east coast of

Marine reptiles and other marine animals were able to use two marine corridors or seaways to move in and out of the Tethys Ocean during the Jurassic. These corridors widened during the Cretaceous as North America moved north and the Gondwanan continents drifted apart.

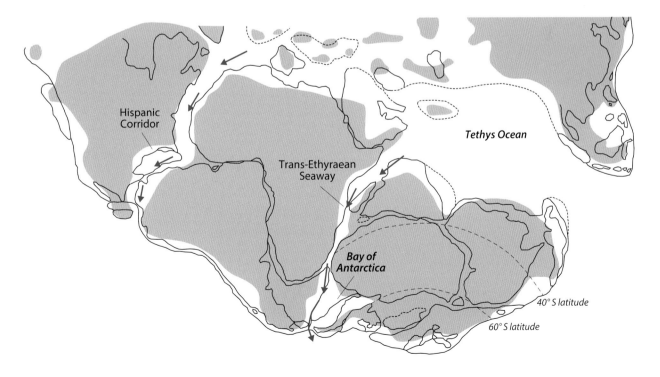

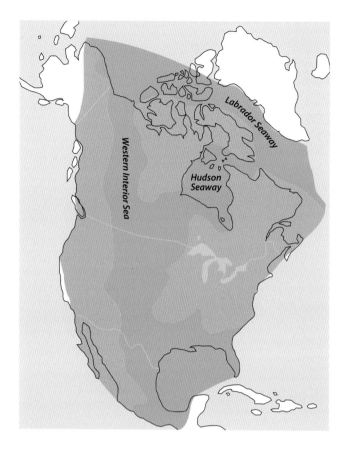

Numerous Cretaceous marine reptile fossils come from sediments laid down in the Western Interior Sea. At its greatest extent, this sea was 3,200 km (2,000 miles) long and 1,000 km (620 miles) wide. It was relatively shallow throughout its history, and generally less than 900 m (3,000 ft) deep.

South America. Much of northern Africa and western South America was underwater during the Late Cretaceous, between 94 and 70 million years ago, due to a rise in sea level.

Throughout the Cretaceous, North America's tectonically active western coast pulled the continent downwards into the crust. Consequently, an arm of the Boreal Ocean moved south, forming the Mowry Sea during the first half of the Cretaceous. An arm of the North Atlantic then moved north from the Gulf of Mexico, and ultimately met the Mowry Sea to form the Western Interior Sea, also called the WIS, Niobraran Sea or North American Inland Sea. North America was divided in two. During the middle of the Cretaceous, an arm of the WIS called the Hudson Seaway connected to the Labrador Seaway west of Greenland, dividing North America into three. The WIS was a key feature of the Late Cretaceous world and was important for mosasaurs and plesiosaurs. Continental uplift ultimately caused the WIS to shrink, but it persisted until the Paleocene, 60 million years ago.

Mesozoic climates and temperatures

Until recently, the assumption about the Mesozoic world was that it was always hot, that the seas were always warm, and that the poles never had ice or snow. Because modern reptiles, especially big ones, are animals of subtropical and tropical places, it made sense that the marine reptiles of the Mesozoic had evolved in such a hothouse world.

Most evidence, including the distribution of fossil organisms and the temperature-sensitive oxygen isotopes preserved within them, shows that while the Mesozoic was warm overall, the world cooled during the Late Jurassic and Cretaceous, and that Late Cretaceous warming made things hotter again. During the warmest parts of the Mesozoic (the Late Triassic, the end of the Early Cretaceous and the early part of the Late Cretaceous), global temperatures were more than 5°C (7.5°F) warmer than those of today, sometimes 10°C (18°F) higher. The oceans at these times had surface waters 10°C (18°F) above the modern average, perhaps almost 20°C (36°F) higher in the tropics.

It is not true, however, that the Mesozoic was perpetually warm. Ice-damaged rock surfaces and stones dropped by floating ice, climatic modelling and isotopes from fossils prove that things were more complex. Cold events occurred during the early part of the Jurassic, in the Late Jurassic, and at least twice during the Early Cretaceous. Some researchers argue that sea and continental ice formed during the Early and Middle Jurassic, and that Western Tethyan sea surface temperatures during these times were close to freezing. Other experts argue that conditions were warmer, perhaps by 10–15°C (18 to 27°F). This still means that many seas of this time were cool, and that animals living here swam in temperate seas, not tropical ones. Actual sea surface temperatures of 10°C (50°F) and less have also been reported for the Early Cretaceous North Atlantic. The seas were also close to – sometimes below – freezing in the far north and south during the Late Cretaceous, yet plesiosaurs and mosasaurs were here at these times.

How did Mesozoic marine reptiles cope with these conditions? It seems that many or most were 'warm-blooded' or endothermic, and able to generate and maintain a high internal temperature. Several lines of evidence support this. The internal texture of their bones shows that they grew quickly (something known to correlate with a high metabolic rate), exceptional fossils show how their bodies were insulated by blubber (meaning they were retaining internally produced heat), and the oxygen isotopes in the phosphate of their teeth are linked to high body temperatures. Endothermy was likely present in ichthyosaurs, plesiosaurs and related sauropterygians, mosasaurs and

It would once have been considered fanciful to show a Mesozoic marine reptile, like this breaching Jurassic pliosaurid, in a sea with ice. However, geological and geochemical evidence demonstrates that cool and cold conditions occurred on several occasions during the Mesozoic. At least some of these animals lived in freezing seas.

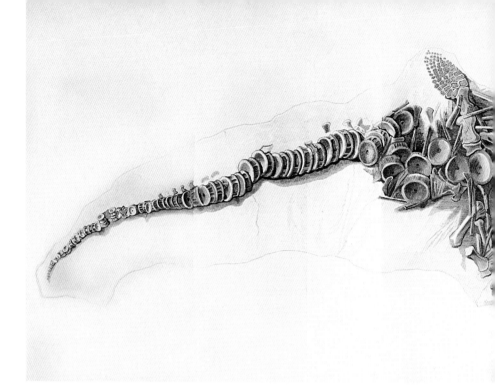

Thomas Birch's ichthyosaur specimen – named *Proteosaurus* by Everard Home in 1819 – was near-complete and significant in revealing the anatomy of the limbs. It was small, at less than 1 m long (3 ft). Sadly, it was destroyed by a bombing raid on London in 1941.

metriorhynchids, which means that it evolved multiple times independently. This explains why several Mesozoic marine reptile groups – parvipelvians and certain plesiosaurs and mosasaurs among them – thrived in cold regions like the Jurassic Boreal Ocean and Cretaceous waters around Antarctica.

MESOZOIC MARINE REPTILE DISCOVERIES

For as long as people have existed, they have found fossils, often realising that they sometimes resemble modern animals. In Europe, ichthyosaur and plesiosaur bones and teeth were identified as those of fish, crocodiles or whales from the 1600s onwards. Some of these were interpreted as vestiges of the biblical flood, such as a plesiosaur from Lincolnshire in England, described in 1719 by archaeology pioneer William Stukeley. A thalattosuchian found in Whitby, England, in 1758 was identified as an alligator or gharial, and a thalattosuchian skull from the vicinity of Roana, Vicenza Province, Veneto, Italy was identified as that of a crocodile in 1794. These were surprising claims given that such animals were known to live far from Europe. Indeed, the people of the late 1700s and early 1800s had yet to learn about extinction, continental movement and evolutionary change.

A particularly significant fossil in the story of marine reptile science was collected from a limestone mine in the Maastricht region of the Netherlands, during the 1770s. It consisted of toothed bones from the jaws and palate of an animal similar in size to a sperm whale. In 1795, Maastricht was invaded by the army of the French Republic, and the specimen was taken by the French to Paris. It is still on display in the Muséum national d'Histoire naturelle.

Most experts thought the fossil was a whale or crocodile, but Dutch mathematician and scientist Adriaan Camper argued it was a gigantic lizard.

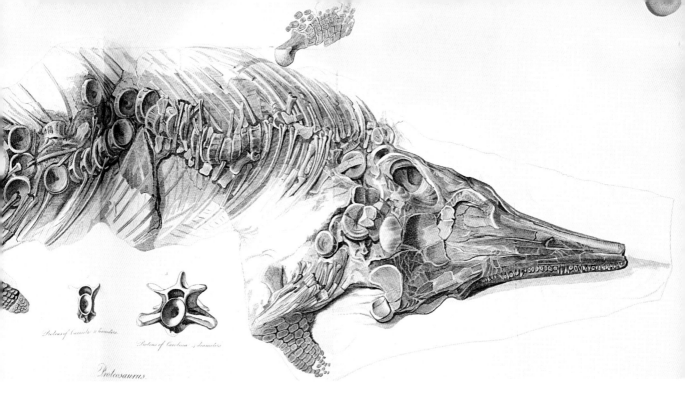

His suggestion was backed by French naturalist Georges Cuvier in 1812, who drew attention to its similarity with monitor lizards. It was named *Mosasaurus*, meaning lizard from the River Meuse, by Anglican cleric and geologist William Conybeare in 1822. No such creature was known to be alive, so here was evidence of an animal that came from the past. This was the genesis of the concept known as extinction. The idea that species die out is familiar today, but it broke new ground in the 1820s. Fossils such as *Mosasaurus* were not just evidence for the existence of previously undiscovered animals, but also a challenge to religious dogma.

At the same time, events in England led to the recognition of ichthyosaurs. In 1811, the Anning family of Lyme Regis discovered a large ichthyosaur in the Lower Jurassic Lias rocks of the Dorset coast. This specimen was described by the British surgeon Sir Everard Home in 1814. Home thought it was allied to fishes, though it appears his ideas were taken from his brother-in-law, the Scottish surgeon John Hunter, who had died in 1793. Other scientists disagreed with Home, instead thinking that ichthyosaurs were reptiles or whales. In 1819, Home concluded that ichthyosaurs were more like salamanders than fishes and he published the name *Proteosaurus*, specifically attaching it to a specimen owned by the British fossil collector Thomas Birch. This name includes a reference to *Proteus*, a cave-dwelling salamander.

Unluckily for Home, the German naturalist Charles König published the name *Ichthyosaurus* for the same animal in 1818, and this is the one that came into widespread use. A complication is that the 1811 specimen that Home first wrote about is not an *Ichthyosaurus*, but a specimen of the larger and very different *Temnodontosaurus*. Its distinction from *Ichthyosaurus* was not properly recognized until 1889.

MARY ANNING

One fossil collector above all others has remained in the public and scientific eye. This is Mary Anning of Lyme Regis in Dorset, southern England. During the time that Mary was alive (1799-1847), Lyme Regis was a holiday resort well known for fossils, including marine reptiles.

Mary's family made their living through furniture making, but also sold fossils to supplement their income. They were poor and often struggled financially. Mary's mother was also named Mary, and some discoveries attributed to the younger Mary were likely made by her mother. Indeed, fossil-finding and selling was a family endeavour. Her brother Joseph, for instance, discovered an ichthyosaur in 1811 when Mary was just 11, and she attributed her knowledge of fossils to the teachings of her father, Richard. After his death in 1810, Mary and Joseph continued the family business.

Victorian interest in Mary Anning promoted the idea that she made her discoveries as a child, but it was while she was in her twenties and thirties that she became famous. Her discoveries included many ichthyosaurs, the first articulated plesiosaur in 1823, the first British pterosaur in 1828, and many

Henry De la Beche's Duria Antiquior ('Ancient Dorset') is one of the first illustrations to reconstruct an ancient ecosystem, and many of the animals it depicts were excavated by Mary Anning. It features many interactions between its animals. Some are eating other animals, and a plesiosaur is defecating (presumably as a fear response) while being grabbed by an ichthyosaur.

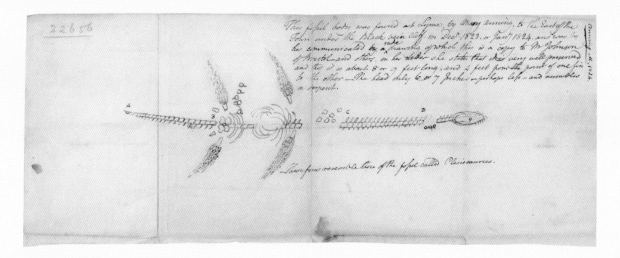

The Anning *Plesiosaurus* specimen caused a sensation when announced in 1824, and experts wrote to each other to share the news. This excerpt of an anonymous letter includes an illustration of the fossil. The illustration might be a copy of a drawing Mary Anning herself produced.

molluscs and fish. She corresponded with numerous scientific experts who became aware of her expertise and even visited her, though the tradition at the time meant that she was generally uncredited for her input and even exploited. As a woman and a working-class person in the provinces she was unable to join or contribute to scientific societies. However, Mary was so successful at finding fossils that she was able to buy a house and a shop with a windowed front in 1826.

Despite claims that Mary has been ignored or forgotten, her fame has been constant since she was alive, and her name is synonymous with fossil hunting at Lyme Regis. Numerous books and a highly fictionalized movie are devoted to her. How much she benefitted from fame when alive is difficult to know, but we do know that scientists and philanthropists of the time helped her financially. In 1830 or thereabouts, British palaeontologist Henry De la Beche produced the illustration *Duria Antiquior* (meaning 'Ancient Dorset'), depicting Early Jurassic life as imagined at the time. The funds raised for its sale were gifted to Mary. She died from breast cancer at the age of 47. Her world-famous contributions to palaeontology are commemorated in the naming of such fossils as *Ichthyosaurus anningae* and the plesiosaur *Anningsaura*. Most people who visit Lyme Regis in search of fossils do so knowing they are following Mary Anning's footsteps, and her legacy was commemorated there with a statue unveiled in 2022.

This portrait of Mary Anning, produced in 1850 and thus after her death, is by B. J. M. Donne, and shows Mary with her dog Tray.

The discovery of *Plesiosaurus*

By the early 1800s, numerous marine reptile fossils were known from the Mesozoic rocks of England, and several collections had been amassed. Views were vague on whether these animals were similar to modern ones such as fish, crocodiles, salamanders, whales or even people, and on how ancient they were. A pivotal study was published by Henry De la Beche and William Conybeare in 1821. They realised that several bones and partial skeletons represented an animal different to *Ichthyosaurus*, and they named it *Plesiosaurus*. This means 'near lizard', chosen because *Plesiosaurus* was 'nearer' conventional reptiles such as crocodiles than the more fish-like *Ichthyosaurus*.

A skull thought to be that of *Plesiosaurus* was described by Conybeare in 1822 (it is today known to belong to another plesiosaur, known as *Stratesaurus*), but more significant was the 1823 discovery of a complete skeleton by the Annings. They sold this to the Duke of Buckingham for between £100 and £200 (equivalent to between £11,000 and £23,000, or $15,000 and $31,350, in modern amounts), and it was then sent to the Geological Society of London. Here, it was discussed at the Society's meeting of February 1824.

Conybeare wrote about the excitement of its arrival in London by sea. Its transit was delayed by uncooperative bad winds, and once at the Geological Society it proved too awkward to get up the stairs. It mostly confirmed Conybeare's thoughts but he had not predicted its remarkable neck, and scepticism about its length was expressed by those unable to see the fossil

The original *Plesiosaurus* skeleton, discovered at Lyme Regis by the Annings in 1823. Mary Anning wrote to William Conybeare about the specimen and included an accurate drawing of its appearance. Consequently, Conybeare had an idea of its completeness and significance before its 1824 arrival in London.

in person. Cuvier, in Paris, asked Conybeare to ensure that the fossil was genuine and not combined from more than one animal, but later finds made it unarguable that this is what plesiosaurs were like. The Anning *Plesiosaurus* was purchased by the British Museum in 1848 and is on display today at the Natural History Museum in London.

Numerous other marine reptile finds were made elsewhere in Europe at about the same time. German Jurassic rocks yielded the thalattosuchian *Geosaurus* in 1816, initially thought to be a lizard similar to *Mosasaurus*, and the gharial-like thalattosuchian *Teleosaurus* was described from France in 1820.

Mosasaurs in Belgium, and news from the USA

The recognition of *Mosasaurus* in the 1770s allowed scientists elsewhere to identify other fossils – including some from the English Chalk – as close relatives. The richness of Europe's mosasaur record was not fully realised, however, until the 1880s and 1890s when Belgian palaeontologist Louis Dollo described a variety of new species from his home country.

In the USA, more new mosasaurs were discovered in the Late Cretaceous rocks of New Jersey, the Gulf Coastal Plain and Kansas Chalk between the 1860s and 1880s, mostly by American palaeontologists Edward Drinker Cope and Othniel Charles Marsh. Some were regarded as American versions of European forms, such as *Mosasaurus* and Dollo's *Plioplatecarpus* and *Prognathodon*. Others were new, such as *Platecarpus*, described from Mississippi by Cope, and the giant *Tylosaurus*, described from Kansas. Cope thought mosasaurs were allies of snakes and gave them a name – Pythonomorpha – reflecting this idea.

Cope published on numerous Late Cretaceous groups beyond mosasaurs, including giant turtles and the short-necked plesiosaurs termed polycotylids. He also reported fossils of another plesiosaur group, this time one famous for having an exceptionally long neck. In 1867, a partial skeleton incorporating over 100 vertebrae was discovered in Kansas. Cope recognized the remains as plesiosaur bones, the most complete from the Americas. He announced the discovery of a new animal that had plate-like limb bones and a long, paddle-shaped tail. He named it *Elasmosaurus platyurus*, meaning 'thin-plate reptile with a flat tail', and produced a skeletal reconstruction in a lavish publication of 1869.

But Cope had made a major error. The 'tail' was in fact the neck, and Cope's short-necked, long-tailed reconstruction was back-to-front. Joseph Leidy – Cope's anatomy professor at the University of Pennsylvania – pointed this out in 1870. Cope immediately tried to recall all copies of his study and have them replaced. Writers have sometimes cast scorn on Cope's misidentification, suggesting he was guilty of a reckless, rushed approach. There is some truth to this, but it's only fair to note that he was working on a radically new fossil where many of the vertebrae were incomplete. Marsh

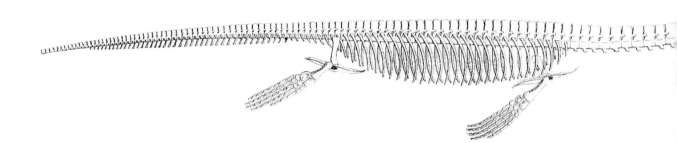

Edward Cope's 1869 reconstruction of the giant plesiosaur *Elasmosaurus* showed it as a remarkably long-tailed animal lacking hind limbs and equipped with a relatively short neck. Propulsion was thought to be provided by sideways movement of the tail.

Cope's revised *Elasmosaurus* of 1870 – hastily published in an effort to correct his incorrect earlier version – at least depicted the long 'tail' as a neck. Cope still thought that the tail might provide thrust during swimming and thought the limbs relatively small and "of comparatively little power".

used this mistake to 'prove' Cope's lack of skill. Both were at loggerheads from 1873 until their deaths in the 1890s, and history has it that Marsh's outing of Cope's *Elasmosaurus* blunder resulted in their becoming enemies. Reality is more complex. The two fell out after a series of conflicts, and Marsh didn't use the *Elasmosaurus* story to 'shame' Cope until 1890, long after they'd fallen out.

New reptiles from the European Mesozoic

The 1800s and early 1900s saw rapid growth in our knowledge of Mesozoic marine reptiles, as discoveries were made across Europe. These bolstered the collections of newly emerging public museums and were studied by pioneering anatomists and palaeontologists, including Georges Cuvier in France, Hermann von Meyer in Germany and Richard Owen and Richard Lydekker in England. Several fossil-bearing rocks proved especially important, including the Triassic Muschelkalk and Posidonia Shale in Germany, and the Jurassic Oxford Clay and Kimmeridge Clay in England.

The Muschelkalk (German for 'mussel-bearing limestone') of central and western Europe dates to the Middle Triassic (237–229 million years ago). Its significance for Mesozoic marine reptile research comes from its sauropterygian diversity, since this is where placodonts, nothosaurs and pistosaurs were discovered. All were known by 1860, a consequence being that a good idea of where plesiosaur ancestry lay was developed relatively early on. Muschelkalk sauropterygians also showed that Triassic Europe was home to a profusion of unusual reptiles that failed to survive into the

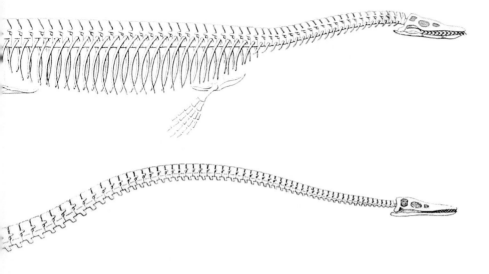

Jurassic. The exceptionally long-necked *Tanystropheus* was also discovered in the Muschelkalk.

Also important is Monte San Giorgio on the Swiss-Italian border in the southern Alps. Monte San Giorgio – a mountain 1 km high – was dug for lamp oil during the 1700s and 1800s, and Triassic fossils were discovered there in the 1840s. During the 1910s, marine reptiles were found, the oldest from Europe. It became obvious that the location was rich in Tethyan marine vertebrates. Hundreds of sauropterygians are known from Monte San Giorgio, including pachypleurosaurs, nothosaurs and placodonts, as are tanystropheids, thalattosaurs and ichthyosaurs. The site was made a UNESCO World Heritage Site in 2003 and is one of the world's best locations for Triassic marine fossils.

Most discussions of Mesozoic marine reptiles mention the Lower Jurassic Posidonia Shale, named for the fossil oyster *Posidonia* (today known as *Bositra*). Exposures of this rock occur across western Europe but are mostly associated with Holzmaden, southern Germany. The Posidonia Shale has been a source of fossils since the 1500s but its Mesozoic marine reptiles were first reported in the 1820s. They include thalattosuchians and plesiosaurs but, most famously, ichthyosaurs preserved with their soft tissues as well as babies within the bodies of their mothers.

Many of these were discovered by the Holzmaden-based Hauff family. Beginning with Bernard Hauff during the 1890s, they discovered, prepared and studied many of the fossils that make the Posidonia Shale famous. During the 1850s, palaeontologist and geologist Friedrich August von Quenstedt determined that the Holzmaden ichthyosaurs were geologically younger than *Ichthyosaurus* from England. In 1904, Otto Jaekel gave them the name *Stenopterygius*, meaning 'narrow wing', a reference to the forefin.

In the English midlands, the Middle Jurassic Oxford Clay is also famous for Mesozoic marine reptiles, the majority discovered during the late 1800s when it was exploited by the brick-making industry. From the 1860s to 1910s, brothers Alfred and Charles Leeds amassed one of the world's greatest Mesozoic marine reptile collections. Vast numbers of invertebrates and fishes have been collected, too, giving us the most complete view of Jurassic marine ecosystems yet compiled. Oxford Clay predators include the parvipelvian *Ophthalmosaurus* and diverse plesiosaurs and thalattosuchians.

A younger English Jurassic rock – the Kimmeridge Clay – has been a rich source of fossils since the 1870s, and yields Mesozoic marine reptiles related to those of the Oxford Clay. Jurassic marine reptile assemblages linked to those of the Oxford Clay and Kimmeridge Clay are also known from Spitsbergen in the Svalbard archipelago north of Norway. Another Upper Jurassic European rock – the Solnhofen Limestone of Bavaria, Germany – is important for Mesozoic marine reptiles, especially thalattosuchians, rhynchocephalians and turtles. The Solnhofen Limestone was quarried during the nineteenth century since its thin, fine-grained layers are ideal for use in printing. A

Monte San Giorgio (the forested mountain visible at right) sits at the south-eastern edge of the sinuous Lake Lugano in Switzerland. The mountain and its surrounds had been quarried for thousands of years prior to the 19th century discovery there of Triassic marine reptiles.

The UK's coastline remains globally important as a source of Mesozoic marine reptile fossils. This is due to their accessibility, their highly active erosional nature and their mostly Jurassic or Cretaceous age. This photo shows a coastal exposure of the Jurassic Oxford Clay near Weymouth, England.

result of this quarrying was the discovery of beautiful, complete fossils, often preserved with their soft tissue outlines. Mesozoic marine reptiles from the Solnhofen region include the thalattosuchian *Geosaurus*, described in 1816 by Samuel von Sömmerring and initially misidentified as a mosasaur.

Jurassic ichthyosaurs and plesiosaurs, again related to those of England, were reported from European Russia during the early 1900s and then throughout the twentieth century. And the Cretaceous Chalk of southern England was also crucial to early views on marine reptile diversity. Mosasaurs were reported from the chalk by pioneering English palaeontologist Gideon Mantell as early as 1829, and the remains of ichthyosaurs and plesiosaurs different from those of the Jurassic were alluded to by Richard Owen, the founder of the Natural History Museum, London, during the 1840s. Here were the first indications that these groups survived much later than previously understood.

A more international scene

Even by the mid-1800s, it was obvious that remains of the Mesozoic marine reptile groups of Europe and the USA would one day be discovered on the other side of the globe. It didn't take long. Australia, for example, yielded ichthyosaurs by the 1860s and remains demonstrating the existence of giant pliosaurids and ichthyosaurs similar to those of Europe were known by the 1920s. New Zealand produced Late Cretaceous mosasaur and plesiosaur fossils during the 1870s, and some of these early finds (like the tylosaurine mosasaur *Taniwhasaurus*) were good enough to show that distinct species were present here.

It was likely that similar animals awaited discovery in Antarctica too, the problem being the difficult fieldwork conditions associated with that remote continent. Not until the 1980s did the far south reveal its first plesiosaurs and mosasaurs. Sure enough, these proved related to those of Australasia and South America. Fragmentary South American plesiosaurs and ichthyosaurs

Mosasaur fossils are mostly associated with North America and western Europe. But African nations are becoming increasingly important as places where mosasaurs are found, Morocco in particular. This undescribed Moroccan specimen (perhaps belonging to *Mosasaurus* or a closely related form) shows how complete these fossils can be.

were discovered during the late 1800s and thalattosuchians in the early 1900s. Better remains from Argentina and Chile were reported from the 1920s onwards. Aristonectine plesiosaurs, platypterygiine ichthyosaurs and metriorhynchids from Chile and Argentina have been described in recent decades and show that specialized lineages were present throughout the south during the later Mesozoic. Northern South America – Colombia in particular – has revealed Mesozoic marine reptile fossils since the 1970s, the region's Cretaceous pliosaurids being especially interesting. Further north, Mexico has been a source of plesiosaurs and thalattosuchians since the 1980s.

African Mesozoic marine reptiles were reported from the early 1900s onwards, including mosasaurs from Libya, Egypt, Angola and Morocco and plesiosaurs from South Africa. Expeditions of the 1980s and 1990s made it clear that Niger, Nigeria and Angola had a richness of unique mosasaurs, many described during the 1990s by palaeontologist Theagarten Lingham-Soliar. The twenty-first century has seen Angola and Morocco yield additional specimens, and the latter is currently a hotspot for new mosasaur species.

Mesozoic marine reptile fossils have been known from the Triassic of China since the 1940s, but a huge number of novel species have been discovered there since the 1990s. Most come from the Middle Triassic shales of Guanling in Guizhou Province and were discovered following road- and bridge-building projects. It turns out that all the Mesozoic marine reptile groups known from the Muschelkalk and Monte San Giorgio are present in China, too. Today, Guizhou is one of the most exciting places for Mesozoic marine reptile discoveries, revealing surprises relevant to behaviour and biology as well as history and diversity.

Mesozoic marine reptile research currently has a more global feel than at any other time, and important discoveries are regularly announced from China, Argentina, Chile, Colombia, Morocco and elsewhere. Despite this, western Europe and the USA remain significant and even Dorset in England and the Holzmaden region continue to produce new species.

The Mesozoic marine reptiles renaissance

Scientists who study dinosaurs are in the middle of what they call the Dinosaur Renaissance, a renewed active research phase. Almost unknown outside of specialist researchers is that a similar event is underway in the world of Mesozoic marine reptiles, and has been since the 1980s. This has seen numerous researchers join the search for new knowledge of these animals. In this case, it isn't that specific discoveries or points of controversy have driven building interest. It is the realisation that a vast amount of work has yet to be done.

For much of the twentieth century, Mesozoic marine reptile research was slow. Influential studies appeared once or twice a decade, and the animals were apparently regarded as uninteresting. A few publications on American plesiosaurs, British and Argentinian ichthyosaurs and German Triassic and Jurassic species kept the subject alive between the 1940s and 1960s. During the 1970s, it became obvious that the European ichthyosaurs and plesiosaurs and American mosasaurs discovered during the 1800s were in need of modern study. Christopher McGowan studied European ichthyosaurs, Beverly Halstead (originally publishing as L B Tarlo) worked on plesiosaurs, and Dale Russell continued work initiated during the 1960s on mosasaurs. These studies inspired much of what happened next.

From the early 1980s onwards, an explosion of studies appeared. The plesiosaurs of the Lias and Oxford Clay became the focus of new study. Jack Callaway, Robert Carroll, Jean-Michel Mazin and others looked anew at marine reptile phylogeny, while Judy Massare produced work on feeding behaviour and biology, and a series of studies on locomotion appeared. Studies devoted to the Triassic fossils of Spitsbergen, Monte San Giorgio and Germany emphasized the role they might play in our knowledge of marine reptile evolution, anatomy and biology.

In the early 1990s, Michael A Taylor highlighted the complexity present in the skulls of large plesiosaurs, his studies being among the first to look at fossil reptile skulls in terms of mechanical performance. A 1993 article by Robert Bakker showed that the view of plesiosaur evolution accepted in previous studies was likely wrong, and that the real pattern was more dynamic. By the mid-1990s, it was obvious that a surge of interest in Mesozoic marine reptiles was underway, the result being the multi-authored 1997 book *Ancient Marine Reptiles*, the first devoted to the subject since Samuel Williston's 1914 *Water Reptiles of the Past and Present*.

Work has continued to expand. A vast number of new species have been reported, innumerable technical studies have appeared on the evolution, anatomy and distribution of these animals, and a good number of scientists – many in the early stages of their careers – publish regularly on the Mesozoic marine reptiles of the UK, Germany, Spitsbergen, Morocco, Angola, China, the USA, Canada, Chile, Antarctica and elsewhere. The field is more active and fast-moving than ever.

2 | EVOLUTION

WORKING OUT WHERE MESOZOIC marine reptiles belong within the reptile family tree has been one of the greatest challenges for scientists interested in these animals. The jaws, teeth, vertebrae and limbs of these animals show that they are reptiles, part of the same group as turtles, lizards, snakes and crocodylians. Most reptiles belong to a group known as the diapsids, more formally Diapsida, characterized by the presence of two openings at the back of the skull, called the temporal openings. The lateral temporal opening is on the side of the skull, behind the eye, while the upper temporal opening is on the skull's upper surface. The lower temporal bar forms the lower margin to the lateral temporal opening, while the upper temporal bar separates the lateral and upper temporal openings. Ancient reptiles from the early part of reptile history lack temporal openings.

Among modern reptiles, turtles lack temporal openings. For this reason, a popular view has been that turtles are not part of Diapsida. A complication here, however, is that we know of many cases where diapsids have modified their skulls such that their temporal openings have become secondarily closed. Might the same have happened to turtles? Genetic studies find turtles to be within Diapsida. Furthermore, ancient turtle relatives from the Permian have temporal openings. Our conclusion, then, is that turtles are diapsids after all, just highly modified ones.

A DIAPSID ORIGIN

The reason this discussion of turtles and their loss of temporal openings is important is that similar, but less extreme, modifications appear to have affected skull shape in some Mesozoic marine reptile groups. Ichthyosaurs and sauropterygians lack lateral temporal openings and have unusual skull bone configurations relative to diapsids, so the 'textbook' view, popular until late in the twentieth century, was that both groups also descended from non-diapsid reptiles.

Ichthyosaurs, it was proposed, perhaps evolved from the same ancient reptiles that gave rise to turtles. Some experts went even further, suggesting that ichthyosaurs were not reptiles, but the descendants of ancient amphibians. This view was supposedly supported by the internal structure of ichthyosaur teeth since these sometimes have a cross-sectional form seen elsewhere in fossil amphibians. Amphibian ancestry was never suggested for sauropterygians, but they were generally included within a group thought to have originated early in reptile evolution. This non-diapsid view of Mesozoic marine reptile origins came apart in the 1980s. Experts studying reptile evolution discovered that sauropterygians in fact possessed characteristic diapsid features, in which case they had to be modified members of this group. If that were true, the absence of the lateral temporal opening in sauropterygians would have to be due to evolutionary modification, and they probably descended from ancestors that possessed it. What seems to have

This family tree shows a possible series of transformations that occurred between early diapsids – like the lizard-shaped Petrolacosaurus from the Carboniferous of the USA – and plesiosaurs. The bones behind the eyes underwent reorganisation, as did the two temporal openings, the position of the nostril, and the form of the teeth.

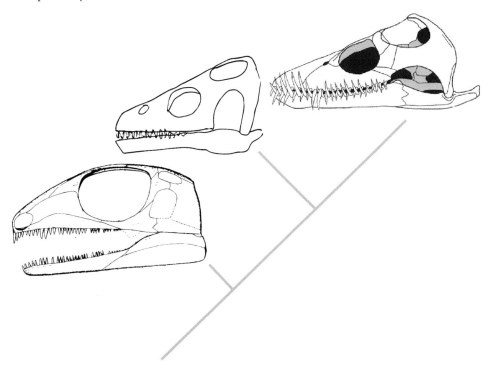

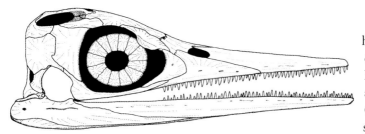

happened is that the lower temporal bar disappeared, the result being that the lateral temporal opening then became an embayment in the cheek region. Over time, the embayment became shallower while the upper temporal bar became deeper and closer to the skull's lower edge.

Triassic ichthyosaur specimens described in the 1990s showed that similar events happened in their evolution, too. Ichthyosaurs were also diapsids that started their history with a lateral temporal opening, but they too lost the lower temporal bar, reduced the size of the embayment, and evolved a thicker upper temporal bar.

If sauropterygians and ichthyosaurs are diapsids, where do they belong within this group? Anatomical features and fossils show that the early diapsids of the Permian, alive 260 million years ago, gave rise to two lineages. One is Lepidosauria, and includes snakes and lizards (united as the squamates), in addition to the rhynchocephalians, a group represented today by the tuatara *Sphenodon punctatus* of New Zealand. Mosasaurs are squamates and thus not of controversial evolutionary position.

The second main diapsid lineage is Archosauromorpha. This includes Archosauria, the 'ruling reptile' group that includes crocodylians and kin (united within Crocodylomorpha), as well as dinosaurs and kin (united within Ornithodira). Archosaurs invaded the seas on several occasions, and one of the groups covered in this book – Thalattosuchia – is among them. Several groups associated with the Triassic are also part of Archosauromorpha, but not Archosauria. They warrant mention here because some of them also invaded the seas, namely protorosaurs and phytosaurs, both of which are covered in Chapter 4.

Since the 1980s, several experts have noted the presence of both lepidosaur and archosauromorph traits in sauropterygians and ichthyosaurs, and affinities with both lineages have been proposed. Also plausible is that sauropterygians and ichthyosaurs evolved from the same ancestors as lepidosaurs and archosauromorphs, but represent a separate, third

This hypothetical ancestral ichthyosaur skull – reconstructed by ichthyosaur experts Michael Maisch and Andreas Matzke in 2002 – shows how ichthyosaurs may have started their evolutionary history with an arch-shaped embayment close to the jaw joint. This embayment is itself a remnant of the lateral temporal opening. Over time, this embayment became shallower until it was lost altogether.

The plesiosaur skull (this is the Jurassic form *Cryptoclidus*) is modified compared to that of an ancestral diapsid. What used to be a lateral temporal opening became an arch-shaped embayment close to the jaw joint. Specialized, interlocking fang-like teeth were also typical of plesiosaurs.

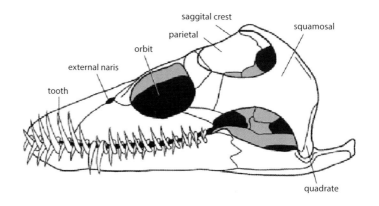

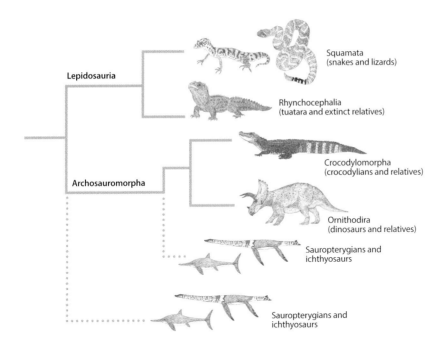

There is little doubt today that Mesozoic marine reptiles are part of the great reptile group Diapsida, the two main branches of which are Lepidosauria and Archosauromorpha. Precisely where Mesozoic marine reptiles belong within Diapsida remains contentious and more than one possibility exists.

lineage. Most recently, a few large-scale studies have found sauropterygians and ichthyosaurs to belong to Archosauromorpha, though well away from protorosaurs and archosaurs.

THE MARINE REPTILE SUPERCLADE HYPOTHESIS

In 2014, ichthyosaur expert Ryosuke Motani and colleagues published a new analysis in their study of the Triassic ichthyosaur-like *Cartorhynchus*. The idea that some Triassic groups – in particular the Chinese hupehsuchians – might be related to ichthyosaurs was suggested in the 1990s, and in fact Motani presented results in 1999 showing that both groups were allied in the clade Ichthyosauromorpha. This 2014 study went further, since its authors found evidence for a clade that includes ichthyosauromorphs, thalattosaurs, the long-jawed *Wumengosaurus* from China, the broad-bodied saurosphargids *and* the sauropterygians. This proposed clade links several Mesozoic marine reptile lineages together and became known as the marine reptile superclade. The possibility that the name Enaliosauria (coined by Richard Owen in 1841) might be attached to the group has been mentioned.

The idea of a superclade is exciting for several reasons. It means that some of the most important Mesozoic marine reptile groups did not start their history as separate lineages, taking to marine life over some drawn-out length of time. Instead, they descended from a single ancestor that became adapted to marine life and quickly gave rise to the founding members of groups that became increasingly disparate. The superclade idea also means that sauropterygians and ichthyosaurs didn't evolve directly from terrestrial ancestors, but from ancestors that had already spent millions of years adapting to marine life.

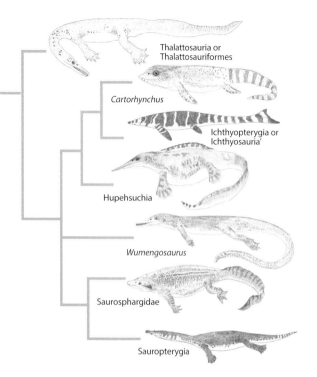

Recent studies suggest that all major Mesozoic marine reptile groups descend from the same single ancestor, and thus form a 'superclade'. Thalattosaurs might be one of the oldest lineages within the superclade. Animals that looked like them seemingly gave rise to ichthyosaurs, sauropterygians and the other groups.

What would the ancestor of the superclade have looked like? Given that thalattosaurs, hupehsuchians, early ichthyosaurs and *Wumengosaurus* share the same basic form, we can infer that this ancestor was about 1 m (3¼ ft) long and had lengthy, slender jaws and a long, laterally compressed tail. Based on the ages of the oldest known members of the superclade, it presumably lived around 255 million years ago, close to the end of the Permian.

TURNOVERS AND EXTINCTIONS

Perhaps only three decades ago, an expert on Mesozoic marine reptiles might have argued that the history of ichthyosaurs, plesiosaurs and thalattosuchians was one of mostly uninterrupted consistency. After emerging in the Late Triassic or Early Jurassic, these animals – so it was thought – underwent little dramatic change. They became bigger, more streamlined and, in the case of plesiosaurs, longer necked, but overall their story was one of minimal change, their world being a stable one, where climate and geography changed little. Copious evidence has overturned this view. Studies on Mesozoic marine reptile evolution have revealed complex histories, and new discoveries have shown how some groups were specialized for life at high latitude, in coastal and estuarine habitats, or dynamic volcanic regions. The history of Mesozoic marine reptiles was punctuated by extinctions when entire groups disappeared, by evolutionary expansions when new groups emerged and quickly took advantage of new environments, and where reorganizations and restructurings of marine ecosystems resulted in some groups dying off while others took on new roles.

The Mesozoic marine reptile story begins with recovery from the greatest mass extinction of all time. This is the End-Permian or Permian-Triassic extinction event of 252 million years ago, and it caused the disappearance of almost 60% of groups alive during the Permian. Its causes mostly involved rising temperatures caused by massive greenhouse gas release. Animal life on land changed as huge numbers of synapsids, the group to which mammals belong, went extinct and were replaced by archosaurs. Life in the seas changed as numerous invertebrate groups disappeared, leaving behind vastly impoverished ecosystems.

Triassic events

Several marine reptile groups were present early in the Triassic, and some, like ichthyosaurs, were represented by species belonging to several lineages.

This suggests that their diversification started in the Late Permian, and is also consistent with rapid recovery from the End-Permian event. Evidence from China and the USA shows that marine ecosystems had mostly recovered within perhaps two million years.

Marine reptile diversity increased during the Triassic. But more extinctions were to come, one in the Middle Triassic 237 million years ago, and another at the end of the Late Triassic 201 million years ago. The second of these – the Triassic-Jurassic extinction event, or Tr–J event – altered plant and animal communities on land and is linked to the rise of dinosaurs. As a consequence, it is well studied. Several factors possibly contributed to the event, including climate change, volcanic eruptions and the impact of extra-terrestrial bodies. The amphibian and reptile groups that dominated terrestrial and freshwater life during the Triassic mostly disappeared, and marine reef communities were hard hit, too.

> The warm, shallow Tethyan environments of the Triassic were home to such marine reptiles as the shellfish-eating placodonts (represented here by *Placodus* on the left) and the more slender, fish-catching thalattosaurs (represented here by *Askeptosaurus*). Alas, environmental changes meant that many of these groups were killed off before the Jurassic began.

Jurassic seas were most obviously home to ichthyosaurs and plesiosaurs, both of which underwent major evolutionary radiations close to the Triassic–Jurassic boundary. This scene shows Early Jurassic leptonectid ichthyosaurs and a rhomaleosaurid plesiosaur, in addition to ray-finned bony fishes and diverse invertebrates.

For marine reptiles, the event 237 million years ago – sometimes called the Ladinian crisis – may have been more devastating. Why it happened is unclear, but global cooling could have been to blame. It meant that only a few groups, including one placodont group and shastasaurian ichthyosaurs, persisted to the end of Triassic times.

Jurassic extinction events

By the start of the Jurassic, the bottom-feeding marine reptiles that were so abundant in the Triassic were gone. Only parvipelvians survived among ichthyosaurs and only plesiosaurs survived among sauropterygians. Both flourished alongside another group: thalattosuchians. The Jurassic saw several temperature and sea level fluctuations, some of which led to prolonged phases of reduced oxygen in the ocean. These are called oceanic anoxic events (OAEs), and one occurred at the start of the Toarcian age near the end of the Early Jurassic, around 183 million years ago. There is no good evidence that this resulted in marine reptile extinction, but a change in marine reptile diversity during the Middle Jurassic could have been a delayed consequence. Several ichthyosaur and plesiosaur groups associated with the Early Jurassic declined or died out at this time.

As the Jurassic ended, 145 million years ago, another event reduced the number of lineages. This is the Jurassic-Cretaceous boundary event. Some experts argue that a major sea level drop occurred at this time, and that outpourings of flood basalt, bolide impacts and global cooling resulted in gradual ecosystem change and the extinction and replacement of animal groups. A reduction in thalattosuchian diversity may be linked to these changes, and several plesiosaur groups declined, too. Ophthalmosaurid ichthyosaurs, however, appear to have been almost unaffected.

Events of the Cretaceous

Early in the Late Cretaceous, around 94 million years ago, another extinction event hit. This was the Cenomanian-Turonian extinction event or boundary event, or C-T extinction for short. It is also known as the Bonarelli event, after Italian geologist Guido Bonarelli, who identified it in 1891. Evidence for the C-T extinction mostly comes from plankton and molluscs, numerous groups of which disappeared. The Cenomanian age was one of fluctuating carbon and oxygen concentrations, and extremely high temperatures and greenhouse gas levels. Climates and environments changed dramatically, with ocean acidity high and oxygenation low. Ultimately, volcanic events in the Caribbean and close to Madagascar might have been to blame. Ophthalmosaurids declined during the Cenomanian until the only ones left at its end were apex predators. Pliosaurids also became extinct at the end of the Cenomanian. The disappearance of both groups opened new opportunities for mosasaurs.

The final Mesozoic extinction event is the catastrophic Cretaceous-Paleogene one of 66 million years ago, usually termed the K-Pg event. This happened at the end of the Maastrichtian, the final age of the Cretaceous, and it eradicated plesiosaurs and mosasaurs. Abundant evidence shows that an object from space, between 10 and 80 km (6 and 50 miles) wide, impacted the Yucatán Peninsula in Mexico, releasing energy equivalent to more than 100 million nuclear bombs and sending vast amounts of rock dust into the atmosphere. Tsunamis kilometres high, wildfires, heat pulses, earthquakes and the vaporising of huge quantities of seawater were among the immediate effects.

In the longer term, ecosystems shut down as plants were unable to photosynthesize, herbivores died, and ultimately predators did, too. Small animals able to rely on rotting plant material, bark or the invertebrates and fungi consuming rotting plant material or bark did survive, but large animals high in the food chain became extinct. A more traditional view proposes that the groups that disappeared were in decline anyway and disappeared gradually due to deteriorating environments. Recent years have seen a massive increase in our knowledge of Maastrichtian plesiosaur and mosasaur diversity, and the opposite was true: both groups were thriving, right to the end of the Maastrichtian, and their disappearance was sudden, not gradual. This is more in keeping with them being killed off by the impact event and its aftermath.

3 ANATOMY

MESOZOIC MARINE REPTILES are so diverse it is difficult to make generalizations about them. Animals built for movement in water are streamlined, have limbs or tails that provide propulsion, sensory organs that enable them to see, smell or hear in water, and jaws specialized for the collection and processing of aquatic food. These statements certainly applied to Mesozoic marine reptiles, as is clear from the shapes of their articulated skeletons.

A few Mesozoic marine reptile groups evolved wide, flattened bodies, sometimes with ridges along the body, including armoured placodonts, cryptoclidid plesiosaurs and sea turtles. These adaptations might have helped provide stability in rough water or channelled water along the body to provide a speed advantage.

Several less obvious but equally important changes affected Mesozoic marine reptile skeletons. The bodies of land-dwelling vertebrates usually have a density similar to that of water. But marine animals that cruise or float at the water's surface have a reduced density. Conversely, some aquatic animals with bottom-feeding lifestyles need to be more dense than water so they can stay low in the water column. Some Mesozoic marine reptile groups – including parvipelvians, thalattosuchians and mosasaurs – evolved spongy, porous bone where the tiny cavities inside were filled with fat, which helped provide buoyancy. The same system later evolved in whales. Bottom-feeding Mesozoic marine reptiles evolved dense, heavy bones. This was done

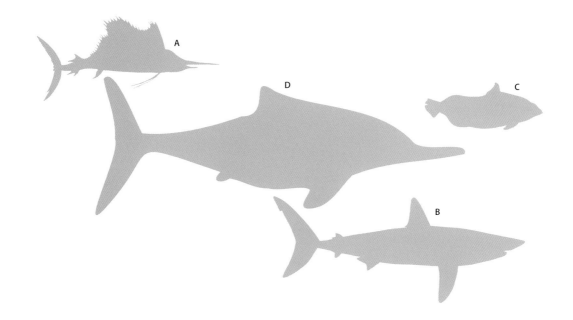

The demands of life in water mean that swimming vertebrates have repeatedly evolved the same streamlined body shape. Fast-swimming ray-finned bony fishes (A), cartilaginous fishes (B), mammals (C) and reptiles like ichthyosaurs (D) are all highly similar in form.

in several different ways, either by growing thicker bone walls, by infilling the bone's interior, or both. Some Mesozoic marine reptiles were dense-boned as juveniles but buoyant-boned as adults, and some had buoyant and dense bones in different parts of the skeleton.

LIMBS, TAILS AND SWIMMING STYLES

An evolutionary trend seen repeatedly in Mesozoic marine reptiles is the change from a system where much of the body and tail are undulated from side to side to one where the body and tail are relatively inflexible, propulsion being provided by the tail's specialized end section. These swimming styles have names based on those of modern fish. The system involving undulation of the body and tail is 'anguilliform swimming', named after *Anguilla*, the eel, while that involving undulation of the tail and not the body is 'carangiform swimming', named after *Caranx*, the jacks and kingfishes. The fastest, most efficient swimming style involves the tail fin alone and is 'thunniform swimming', after *Thunnus*, the tunas. A transition from anguilliform to carangiform and ultimately to thunniform swimming occurred in ichthyosaurs, thalattosuchians and mosasaurs as they took to life in the open ocean (properly known as the pelagic environment), evolved larger size and became speedy cruisers adapted to cover distance at increased efficiency. All evolved a vertical tail fin formed of modified vertebrae and internal stiffening structures.

Sauropterygians evolved from undulatory swimmers and the tail likely provided some propulsion in most groups. However, they also developed

Three main swimming styles – (left to right) anguilliform, carangiform and thunniform swimming – are seen in fish, and also in reptiles that returned to aquatic life. The styles differ in how much body and tail movement is involved, and also how speedy and efficient they are.

A change in the angle of the vertebral column and in the shape of the vertebrae (opposite) mark the location of a tail fin in those marine reptiles that have one. Rarely, the fin is preserved as an impression or trace. From top to bottom: preserved tail fins in a mosasaur, metriorhynchid and parvipelvian ichthyosaur.

a modified pectoral girdle and would have used the limbs, too. A trend whereby the tail became smaller and less important while the limbs became the main propulsive organs occurred in the lineage leading to plesiosaurs. There is so much to say about their swimming behaviour that this is covered separately in Chapter 6 (see pp. 129–130).

Dorsal fins evolved in ichthyosaurs, presumably to maintain stability during fast swimming. There is no evidence that other groups such as thalattosuchians and mosasaurs possessed a dorsal fin, though the possibility that they had one has been suggested.

TEETH AND FEEDING

Mesozoic marine reptiles tend to have conical teeth suited for grabbing fish and invertebrates. In many of these animals, the teeth are set in sockets, the condition termed thecodonty. Ichthyosaur teeth are usually implanted within grooves that extend along the jaws, a condition termed aulacodonty. Teeth are normally distributed along the edges of the jaws, but reptiles started their history with teeth on the palate too, and this was retained in placodonts and in certain ichthyosaurs and mosasaurs. Several Mesozoic marine reptile groups, including placodonts, ichthyosaurs and mosasaurs, evolved globular teeth, and placodonts evolved large, flattened teeth too.

These were built for diets involving shelled molluscs and crustaceans. Fine, closely spaced teeth used in sieving and filtering evolved a few times, including in the specialized aristonectine plesiosaurs.

Reptiles usually undergo tooth replacement throughout life, where replacement teeth ordinarily sit beneath the roots of functional teeth and push them out as they grow. Such was the case in Mesozoic marine reptiles. Sauropterygians are unusual because their replacement teeth grew within chambers located on the inner side, the tongue side, of the functional teeth. As the replacement teeth reached maturity, they migrated sideways – towards the outer side of the face and lower jaw – into the tooth socket. Growth lines within teeth show how quickly they grew and were replaced. One study has shown that tooth replacement in elasmosaurid plesiosaurs was slow, individual teeth lasting two or three years. Perhaps this is because a precisely interlocking, symmetrical tooth arrangement was needed for successful prey capture.

American palaeontologist Judy Massare published an important study in marine reptile feeding behaviour in 1987. Massare devised a way of classifying teeth according to their proportions, the shapes of their tips, their textures, and their breakage and polish. She found that teeth could be classified into seven groups, termed guilds. This term is used to describe a group of animals that use resources in the same way. 'Crush' guild animals have enlarged teeth built for the crushing of shelled prey, while 'crunch' guild

These diagrams show cross-sections through the upper and lower jaws of an elasmosaurid plesiosaur. A replacement tooth (blue) began growth on the inner part of the jaw before moving sideways and into the socket occupied by the older tooth (red), which it eventually replaced.

species have robust, blunt teeth used to disable armoured prey. Animals in the 'smash' guild have small teeth with rounded tips, built to grasp soft prey such as squid, while 'pierce' guild animals either have slender, sharply pointed teeth used to pierce prey ('pierce I') or more robust but still pointed teeth, often with broken or worn tips ('pierce II'). 'General' guild animals have pointed teeth that bear numerous ridges and look suited for handling a variety of prey bigger than those captured by 'pierce' guild animals. Finally, 'cut' guild animals have thick, pointed teeth equipped with cutting edges, built to kill and dismember large vertebrates.

Stomach contents discovered within fossils reveal the prey some Mesozoic marine reptiles ate prior to death, and these data mostly match Massare's correlations. There are, however, indications that these correlations do not always apply. An example is provided by Triassic ichthyosaurs which have 'pierce' guild teeth yet were consuming large vertebrates.

> Pliosaurid teeth like these are typically conical and curved along the length of the crown. The crown is typically about 10 cm (4 in) long. Vertical striations – a common feature on the teeth of big, predatory marine reptiles – probably helped the teeth break through the tissues of prey. These teeth are from the Jurassic pliosaurid *Liopleurodon*.

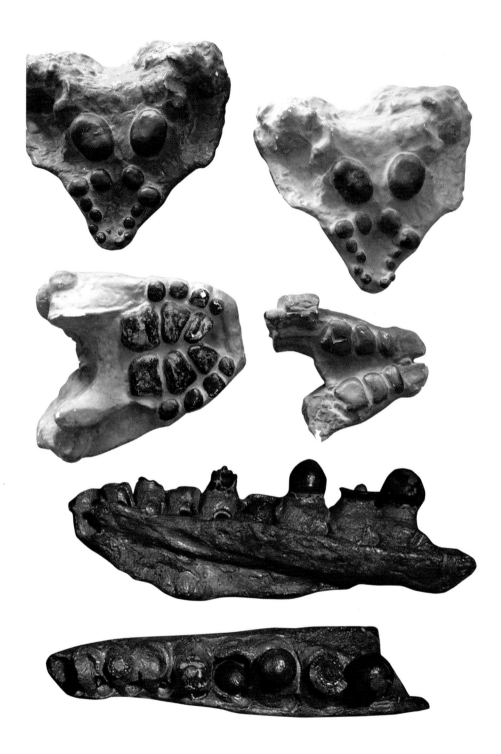

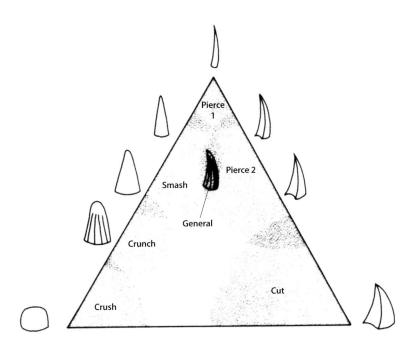

This diagram is from Judy Massare's 1987 classification of tooth types in Mesozoic marine reptiles, a seminal study of marine reptile ecology and behaviour. The categories Massare identified are not entirely distinct but have some degree of overlap. Teeth suited for the piercing of prey, for example, can also be suited for cutting the bodies of large prey items.

Teeth (opposite) specialized for crushing or breaking shells evolved several times in Mesozoic marine reptiles. During the Triassic, placodonts (shown at the top) possessed both rounded and flattened teeth on both the jaw edges and the palate. Tens of millions of years later, during the Cretaceous, rounded teeth also evolved in certain mosasaurs.

THE SKULL

Several generalizations apply to Mesozoic marine reptile skulls. The eyes are large and the nostrils high on the snout and often close to the eyes. The latter is not the case in thalattosuchians where the nostrils are close to the snout's tip. Special salt-excreting glands, required to excrete the excess salt consumed as a consequence of eating marine prey, must have been present in or on the Mesozoic marine reptile head just as they are in living sea turtles, seabirds and sea snakes, and different groups evolved these glands in different locations.

Marine reptile temporal openings are large, sometimes enormous relative to the size of the skull. In plesiosaurs, a midline crest is sometimes present along the rear of the skull, the bony bars around the eyes and temporal openings are often broad and deep, and a backward-projecting bony bar called the retroarticular process is usually present at the rear of the lower jaw. These features are linked to jaw musculature. They show that the jaw muscles were usually huge, that the jaws could be opened and closed quickly, and that many plesiosaurs bit with tremendous force.

Mosasaurs have a notably open skull structure, due to their squamate ancestry, where the side of the skull behind the eyes forms a series of bars and arches. Squamates possess hinges and joints between the skull roof and snout, between the braincase and palate, in bones close to the jaw joint, and midway along the lower jaw. These features mean that squamates have cranial kinesis - the ability to flex, expand, widen and rotate sections of the skull when opening or closing the jaws. Most mosasaurs retained these features and were able to widen the jaw when swallowing or move the muzzle up and down when handling prey.

Sauropterygians have a well-developed bony palate which underlies the braincase at the back of the skull. This was presumably the anchor site for a flap that kept the windpipe sealed when water was in the mouth, but it might also have helped in holding and crushing prey. The plesiosaur palate possesses openings along its midline and sides, among which are paired openings about half-way along its length, identified by some experts as internal nostrils. In some plesiosaurs these are V-shaped and connected to grooves that extend forward to the snout's edges. Inside the snout, the internal nostrils are connected by tubes to the external nostrils on the outside of the skull. Both the internal and external nostrils are surrounded by hollows that create the impression that water flowed in or out of them.

In 1991, British palaeontologist Arthur Cruickshank and colleagues argued that plesiosaurs used 'hydrodynamically driven underwater olfaction'. The shape and location of the internal nostrils, they proposed, allowed water to flow into the snout chamber where scent particles could then be detected. The external nostrils, they suggested, were not used in breathing, but in the expelling of that water. Breathing must instead have been done via the mouth. Cruickshank and his colleagues proposed this model following their study of the plesiosaur *Atychodracon* from the Early Jurassic. They implied that it was present across all plesiosaurs, and maybe other sauropterygians and ichthyosaurs, too. However, their idea has been contested by other experts, some of whom point out that the internal nostrils crucial to the model might be openings of a different sort, the true internal nostrils being further back on the palate. If this is so, plesiosaurs might have used their external nostrils in ordinary fashion (that is, in breathing) and the concept of 'hydrodynamically driven underwater olfaction' might be bogus. This is one of many areas in marine reptile biology where experts disagree, and where only further research or discoveries will enable us to work out which option is correct.

Ordinarily in reptiles, the pillar-shaped quadrate bone at the back of the skull has a notch on its rear edge for the ear drum or tympanum. A rod-shaped bone called the stapes is connected to the quadrate and conducts sound from the tympanum to the inner ear. Because the tympanum evolved to transmit airborne sounds, it is of limited use in water and Mesozoic marine reptiles tended to lose it over time. In sauropterygians, pachypleurosaurs had a tympanum while at least some plesiosaurs did not. Some plesiosaurs fused the stapes to the surrounding bones and some lacked it entirely. A slender stapes that appears consistent with hearing airborne sounds is known for the Early Jurassic plesiosaur *Microcleidus*, so the group may have been diverse in ear anatomy and hearing behaviour.

Mosasaurs modified the quadrate so it became a massive, thick, bowl-shaped bone. It appears to incorporate a calcified tympanum. Some experts have suggested that the quadrate was like this to protect the tympanum from damage, and to be linked to a diving lifestyle where changes in pressure

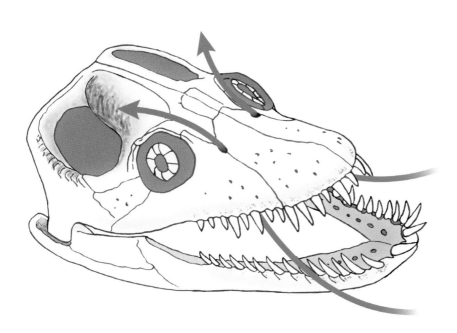

The idea that plesiosaurs used their nostrils as part of an 'underwater sniffing system' was put forward in 1991. This might explain several unusual features of plesiosaur palate and nostril anatomy. However, the proposal is controversial and some experts argue that a more 'normal' function for the nostrils should be favoured instead.

might otherwise have resulted in ear damage. However, specimens where the soft tissues are preserved show that there was no external ear opening, so the modified tympanum was submerged in skin and likely did not need protection anyway. The fact that it was covered also shows that these animals were not relying on airborne sounds.

On that note, there is no reason to assume from these ear modifications that Mesozoic marine reptiles were deaf. Many modern animals that lack external ears can still detect sound and some, like certain frogs and snakes, use sound to communicate. The sounds are transmitted to the inner ear and brain via the mouth or even the whole body. After all, sound waves travel through the bodies of animals just as they do through water, the only problem being that the source of the sound cannot be pinpointed. It is possible that marine reptiles of some or many sorts generated noise via the voicebox, or larynx, and via flipper slaps, jaw snaps and so on.

Like all vertebrate animals, Mesozoic marine reptiles possessed fluid-filled tubes – termed semi-circular canals – within their inner ears. These would have given them their sense of balance. Semi-circular canal anatomy has been studied in sauropterygians and much variation exists, plesiosaurs possessing an unusual anatomy that differs from that of Triassic groups such as placodonts. The plesiosaur configuration is similar to that of sea turtles, and the placodont one that of crocodiles, so these findings are in keeping with ideas on the lifestyles of these groups. Placodonts were coastal animals of the shallows while many plesiosaurs were pelagic.

THE VERTEBRAL COLUMN

The bones that form an animal's spine are called vertebrae (singular - vertebra). These vary according to their position. Vertebrae from the body are more closely locked together than those from the neck and tail and have articulatory surfaces, or facets, on their sides to allow movement of the ribs. In general, the vertebrae of Mesozoic marine reptiles articulate with one another in a way that allows some side-to-side movement. However, tightly articulated vertebrae allowing little movement are seen in the necks, bodies and tail bases of parvipelvians and some mosasaurs. Slender structures termed neural spines grow upwards from vertebrae and provide attachment points for the back muscles. Bony rods termed chevrons serve a similar function along the underside of the tail.

Ordinarily in reptiles, vertebrae from the hip region form a structure termed the sacrum, the function of which is to transmit weight (via bony bars termed sacral ribs) from the hindlimbs, through the pelvic girdle, to the spine. Aquatic animals do not need to transfer weight in this way, so Mesozoic marine reptiles tend to have a sacrum early in their evolutionary history but only the relics of one later on.

> The down-curved end section of Jurassic ichthyosaur tails led to suggestions that a fleshy fin was present here. Experts like Richard Owen suspected that this fin was low and squared-off in shape, rather than tall and triangular. This look is captured in numerous reconstructions of the time, including this one (the ichthyosaur is the large animal on the right).

Ichthyosaur skeletons from the Lias tend to have a downward bend at the end of the tail, making it look as if the spine was broken. In 1838, Richard Owen suggested this was linked to the presence of a fleshy fin that caused the bones to dislocate after death. Specimens from Holzmaden in Germany uncovered in 1892 revealed the true arrangement - the vertebrae extended downwards to support the lower lobe of a crescent-shaped tail fin shaped like that of modern thunniform swimmers. A similar down-turned tail section was later noted in metriorhynchids and certain mosasaurs. Did they have tail fins, too? The answer is yes, as revealed by specimens also preserved with soft tissue outlines.

Plesiosaurs possessed a short tail that was neither especially mobile nor important in swimming. A soft tissue outline preserved around the Mexican *Mauriciosaurus* shows that the tail was broad-based and triangular when seen from above or below. The vertebrae at the tip of the plesiosaur tail are usually compressed from side to side, sometimes fused together and with neural spines that project at a different angle relative to those elsewhere. These features seem linked to the presence of a fin and it seems that tail fins were a typical plesiosaurian feature, differing in size and shape from group to group. One fossil – the original specimen of *Seeleyosaurus*, described in 1895 – appears to have a diamond-shaped tail fin. In aristonectine plesiosaurs, the vertebrae at the tail tip are flattened in the vertical plane, indicating that a horizontal fin was present.

THE LIMB GIRDLES, CHEST BONES AND GASTRALIA

Reptiles ordinarily possess a series of bones in the shoulder region that form the socket for the forelimb and anchor various forelimb muscles. This is the pectoral girdle, and its biggest bones are the shoulder blade or scapula (plural - scapulae) and coracoid, both of which are at the front of the ribcage. A similar series of bones form the pelvic girdle close to the junction of the body and tail. Here, the ilium is at the top while the pubis and ischium support the sides and underside of the gut. The limb girdles of Mesozoic marine reptiles no longer had their original, weight-bearing roles, and in most groups diminished in size.

Plesiosaurs have one of the most unusual pectoral girdles of all. Their scapulae descended on the side of the ribcage and developed new sections directed toward the chest's midline. In most plesiosaurs, the scapulae were large plates, mostly underlying the ribcage and meeting along its midline. The clavicles, or collarbones, existed as reduced midline remnants located at the leading edge of the girdle. Meanwhile, the coracoids were giant plates with a long midline contact along the chest's underside. This uniquely enlarged pectoral girdle provided the attachment site for muscles used in pulling the wing-like forelimbs up and down.

THE PLESIOSAUR NECK

If there's one thing that plesiosaurs are known for, it's a long neck. In contrast to the mammal neck, which virtually always has seven vertebrae, the plesiosaur neck is variable in its number of vertebrae, which ranges from 12 to 75. The long necks of elasmosaurids look serpent-like, and a consequence is that nineteenth-century experts assumed plesiosaur necks to be exceptionally flexible, and able to form coils or s-shapes or bend far to the side or up and down. However, necks and the vertebrae that form them are complex. A single vertebra consists of the cylindrical body, known as the centrum (plural - centra), and the neural arch on top of it. The arch houses the spinal cord (it runs through the arch), is topped by the upward-projecting neural spine, and has interlocking structures termed zygapophyses at both ends. In elasmosaurids, the zygapophyses at the rear form a U-shaped notch while those at the front form a rod that fits into the notch of the preceding vertebra. The union between these structures clearly allowed some movement, but some experts have argued that it was built to keep the neck rigid. It has also been argued that plesiosaur neck vertebrae are so tightly articulated that there was little chance of movement between them. For these reasons, some experts have argued that the elasmosaurid neck was mostly inflexible, and that the main advantage it provided was in allowing the animal to reach down to the seafloor without being close to it. Contradicting this is the fact that the necks of live plesiosaurs would have included cartilaginous pads between the vertebrae. Even if only a few millimetres thick, these would have increased the motion possible at every joint along the neck. In a neck consisting of more than 30 vertebrae, there only needs to be a few degrees of mobility at each joint for the whole neck to bend a great deal.

In 2008, Australian palaeontologist Maria Zammit and colleagues built model plesiosaur neck vertebrae which they articulated, making sure the motion was consistent with the presence of intact cartilage. They found that the neck could bend in every direction, including nearly 180° to the side, and in S-like horizontal curves, but not vertical ones. Several fossil plesiosaur skeletons have their necks bent in poses consistent with these conclusions. Studies have also examined the hydrodynamic effects of the plesiosaur neck. Bending the long neck while swimming would have had a braking effect and could even have been damaging. What this suggests is that major neck movements probably did not happen during rapid swimming, and that the neck was held straight when the animal was moving forward at speed. As a generalization, however, plesiosaur necks were not stiff or immobile, but nor were they as bendy and snake-like as shown in some reconstructions.

Plesiosaurs had so many neck vertebrae that even a small amount of bending at each intervertebral joint would result in a substantial amount of neck flexibility. These diagrams (showing an elasmosaurid) depict the minimum amount of flexibility in both the vertical and lateral planes.

Ichthyosaurs also have a robust pectoral girdle, the T-shaped interclavicle forming a connection between the bar formed of the clavicles at the front of the chest and the plate-like coracoids further back. At the edges of the ribcage, both the clavicles and coracoids are connected to the scapulae on the ribcage's outer sides. The mosasaur pectoral girdle features a fan-shaped scapula and a large coracoid that mirrors the scapula in shape, the bone flaring outwards from a narrow base. Rough edges around these bones show that they were expanded by cartilage, which would have kept them connected and allowed for flexion and bending.

These pectoral girdles are sufficiently robust and muscular that experts have suggested they might indicate a special function. In 1986, German ichthyosaur expert Jürgen Riess argued that ichthyosaurs used their forelimbs in underwater flight, a surprising claim in view of ichthyosaur tail anatomy. The argument that some mosasaurs might have been underwater fliers has also been made. In neither case do the animals possess the large, wing-shaped flippers required for underwater flight to work, however. What seems instead to be the case is that the large and powerful pectoral girdles of these groups were linked to the use of their forelimbs in braking, turning and accelerating.

Because the pelvic girdle encircles the exit for the gut and reproductive tract, its size and shape can provide insight on whether fossil reptiles laid eggs or gave birth. A few studies have looked at pelvis shape in Mesozoic marine reptiles and the results are consistent with viviparity – the ability to give birth to live young – in thalattosuchians and in at least some Triassic sauropterygians.

The underside of the body in reptiles typically involves a breastbone or sternum, located further back than the coracoids and connected to the ribs by cartilaginous sternal ribs. Sternal ribs are preserved in some mosasaurs where they look like those of living lizards, but their distribution in other Mesozoic marine reptile groups is less clear. They appear absent in plesiosaurs, as does a sternum. Located between the sternal region and the front of the pelvis are interlocking bones called gastralia or belly ribs. These are absent in mammals, and it's partly for this reason that people have tended to consider them an unusual feature, whose existence should be 'justified' by reference to some aspect of biology. For many reptile groups, gastralia are present simply because they were inherited from their ancestors. In plesiosaurs, however, the gastralia are large, thick and dense. This suggests that they played an important role, perhaps in keeping the body rigid. They could also have helped in buoyancy by adding mass and helping the animal remain at depth.

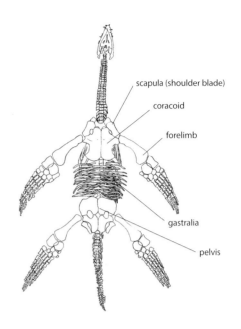

Seen from underneath, the plesiosaur skeleton features the massive, flattened bones of the pectoral girdle (connected to the forelimbs) and pelvic girdle (connected to the hind limbs). A series of interlocking 'belly ribs' or gastralia form the so-called 'gastral basket' between the limb girdles.

LIMBS AND THEIR FUNCTION

For marine animals to move efficiently, their limbs have to be streamlined paddles or flippers good at cutting through water and generating thrust. Because Mesozoic marine reptiles descended from terrestrial ancestors equipped with limbs built to support weight on land, the same evolutionary changes happened again and again as they took to the water: the number of mobile joints in the elbow, wrist, hands and fingers was reduced as the limb became a unified, semi-rigid organ; meanwhile, the long, roughly cylindrical bones that formed the lower arm (radius and ulna), lower leg (tibia and fibula), hand (the metacarpals) and foot (the metatarsals) became shortened, block-like and increasingly similar in size and shape; and claws became reduced and lost since their main function was to secure purchase on land.

The results of these changes are limbs where the bones no longer look like their ancestral versions. The limb bones of many ichthyosaurs and plesiosaurs for example, look (apart from the humerus and femur) like polygons, rectangles or ovals, sometimes tightly fitted together and sometimes arranged loosely.

In concert with this trend was another in which the bones of the fingers and toes – termed phalanges – underwent duplication, presumably because this made the limbs longer and more effective in swimming. From ancestors with three, four or five phalanges per digit, Mesozoic marine reptiles evolved a condition called hyperphalangy, where the digits have more than 10 and even 20 or 30 phalanges. In some groups, including early ichthyosaurs and most plesiosaurs and mosasaurs, the phalanges are cylindrical or dumbbell-shaped, while in others – such as parvipelvians – they are rounded or oval.

The small phalanges at the tips of digits tend to separate from the rest of the limb during decomposition, and as a consequence we are often unsure of the exact shape of the limb's end in some species. Nevertheless, it is obvious that some groups – including ichthyosaurs like temnodontosaurs and leptonectids, virtually all plesiosaurs and at least some mosasaurs – possessed long, tapering limbs, while others – like some ophthalmosaurid ichthyosaurs – had shorter, broader, more rounded limbs. These differences are linked with different swimming techniques and lifestyles - those groups with long, tapering limbs used them as underwater wings or steering devices, while those with shorter, broader limbs used them more in small scale manoeuvres and braking only.

CARTILAGE, SKIN AND OTHER SOFT TISSUES

Vertebrate animals are not made of bones alone. Cartilage is present at the joints and sometimes forms an extension around a bone or part of it. Because this tough yet flexible material ordinarily lacks mineral crystals, it tends not to be preserved in fossils, though there are cases where it is.

Mesozoic marine reptiles tend to have substantial amounts of cartilage in their limb joints and limbs, mostly because their skeletons did not need to be as extensively ossified, or bony, as those of land-living reptiles. This is a result of the reduced gravitational demands of living in water. A consequence of having a cartilage-heavy skeleton is that it did not change as much with age as it did in land-living reptiles. Cartilage is a feature of the juvenile skeleton, so many Mesozoic marine reptiles retained 'juvenile' features into adulthood, a phenomenon called paedomorphosis.

Finally, neither internal organs nor the skin tend to be preserved in fossils, since they are often broken down by scavengers and bacteria long before the body becomes entombed in sediment. This means that we usually lack information on what the organs were like *and* on the animal's external appearance. There are, however, occasional cases in which oxygen-poor bottom waters and rapid burial led to the preservation of organs and skin.

As streamlined, aquatic animals, it seems plausible that the skin of Mesozoic marine reptiles was smooth. However, they are reptiles, so a scaly covering should be assumed. Numerous Jurassic parvipelvians are preserved with a skin outline and even patches of skin intact. Certain of these fossils, most famously those from the Holzmaden region, demonstrate that ichthyosaurs had a triangular dorsal fin and vertical tail fin. Several *Stenopterygius* specimens and a *Mixosaurus* reported in 2020 prove the presence of scaleless skin in ichthyosaurs. It is assumed that this was the condition throughout ichthyosaurs, though whether this was also true of early members of the ichthyosaur clade (like hupehsuchians and nasorostrans) is unknown.

Some ichthyosaur skin sections reveal layers of internal skin fibres. These are arranged in a mesh-like fashion as they are in living vertebrates specialized for fast swimming, and keep the skin smooth and streamlined. One *Stenopterygius* specimen shows that this animal had a thick layer of fat (typically termed blubber in aquatic animals) several centimetres (a few inches) thick. We might have already assumed this, given the need for these animals to be streamlined and retain body heat. It appears that plesiosaurs and mosasaurs were insulated by blubber too, just as leatherback turtles are today. Many of these animals would have been 'plumper' than they appear in traditional reconstructions, and plesiosaurs likely had deeper, thicker necks than what is conventionally shown.

Ichthyosaur soft tissues show that their limbs were complex, with internal fibres and plate-like structures surrounding the bones

These parallel fibres – each around 0.1 mm (0.004 in) thick – can be seen in the skin of some well-preserved ichthyosaur specimens. They appear to be collagen fibres, arranged diagonally in overlapping layers. The fibres in other layers were thicker, the thickest being 0.5 mm (0.02 in) thick.

and reinforcing the edges and tips. These fossils show that parvipelvian fins were broad-based at their point of contact with the body. A few plesiosaur fins preserve their soft tissue outlines. They show that the trailing edge extended beyond the bones, that a curved, flexible tip was present, and that the limb's base was narrower than its main blade. The limb was similar in form to a penguin or sea turtle flipper, an observation consistent with studies that suggest plesiosaurs swam with a flapping motion.

Several fossils shed light on thalattosuchian skin. Teleosauroids had scaly skin like that of crocodylians, while metriorhynchids lacked scales entirely. The condition in mosasaur has been known since 1878, when patches of scaly skin were described for *Tylosaurus* from Kansas. In fact, mosasaur skin is known from five species, representing most major branches of the mosasaur family tree. They show that mosasaur scales were tiny (a few millimetres long, or less than a tenth of an inch), diamond-shaped, and typically with a raised ridge along the middle. The possibility that some mosasaurs – perhaps specialized ones from late in the group's history – might have lacked scales has been suggested, but is at odds with evidence found so far. A few mosasaur specimens have overlapping fibre bundles preserved within the skin. As in ichthyosaurs, these probably helped keep the skin rigid during swimming.

Reconstructions of mosasaurs have often shown them with a frill on the neck, back and tail. This was included in Charles Knight's artwork of

the 1800s, and consequently in toys and films. However, by the 1890s it was known that this interpretation was a mistake, and that the 'frill' was a misidentified, misplaced windpipe or trachea.

Finally, one detail of Mesozoic marine reptile appearance we thought we would never know is colour. Based on the pigmentation of large fishes and marine mammals, it had been assumed that parvipelvians, plesiosaurs and mosasaurs were grey, blue or black, perhaps with a pale underside. Spots and stripes that might help break up the outline seem likely in view of how common they are in marine animals today. These speculations still apply, but recent years have seen the discovery of fossilized pigments in Mesozoic marine reptile skin. *Stenopterygius* specimens from England and Germany preserve cells indicating that the skin was entirely black, including on the underside, though one specimen appears to have had a pale belly. Maybe *Stenopterygius* changed their pigmentation during life, or maybe these fossils belong to different species. Evidence for black skin has also been reported for the mosasaur *Tylosaurus*. This does not mean that all mosasaurs were like this and it remains plausible that complex patterns, bright colours and prominent countershading was present. Other Mesozoic marine reptiles may, similarly, have been diverse and complex in pattern and colour, these things varying according to their habitats, lifestyles and behaviours.

Several paintings by the American palaeoartist Charles Knight depict mosasaurs like this *Tylosaurus* with an elaborate, wavy-edged frill. Knight was incorporating information gleaned from fossils and which appeared to preserve this surprising detail. Alas, it was eventually shown that the preserved 'frill' was no such thing.

4 | THE LESSER-KNOWN GROUPS:
MESOSAURS, TRIASSIC SAUROPTERYGIANS, CRETACEOUS SEA SNAKES AND MORE

Life in the Late Triassic seas of Guanling, southern China. The ichthyosaurs *Guanlingsaurus* (background) and *Qianichthyosaurus* (foreground) lived alongside thalattosaurs (like *Miodentosaurus*, at middle right), placodonts, early turtles and diverse fishes. The plant-like objects are sea lilies or crinoids, distant relatives of sea urchins. They formed massive rafts attached to floating wood.

THE TERM MESOZOIC MARINE REPTILE usually refers to the main characters of this book: the ichthyosaurs, plesiosaurs, thalattosuchians and mosasaurs. However, living alongside or preceding these are a larger number of lesser known, less species-rich groups, which are best characterized as being obscure and weird. Most are associated with the Triassic, but a few were alive during the Jurassic and Cretaceous. Also relevant to our story are animals that are not from the Mesozoic at all, but lived during the Permian, the last part of the Paleozoic, the geological era before the Mesozoic. While technically beyond the remit of this book, they require inclusion because they might be directly connected to the Mesozoic groups.

These groups from the Permian include the mesosaurs and claudiosaurs. The Triassic ones are the placodonts, pachypleurosaurs, nothosaurs, pistosaurs, protorosaurs, phytosaurs, hupehsuchians, thalattosaurs, saurosphargids and helveticosaurs. Several of these are close relatives of plesiosaurs and belong with them in Sauropterygia. Others are close relatives of sauropterygians and belong with them within the superclade (see Chapter 2). A few Triassic marine reptiles are not part of the superclade but do belong to Archosauromorpha, the reptile clade we met in Chapter 2. They include the protorosaurs and superficially crocodile-like phytosaurs.

The Jurassic and Cretaceous groups included in this chapter belong to Lepidosauria, the group that includes lizards, snakes and their relatives. They

are the swimming rhynchocephalians – a group that have no evolutionary connection with any of the other marine reptile groups of the Mesozoic – and the sea-going pachyophiid or simoliophiid snakes. These might, some experts argue, have an evolutionary connection to mosasaurs.

What the groups in this chapter show is that there was constant opportunity for reptiles to take to marine life during the Mesozoic, something we might predict given the nature of the Mesozoic world (see pp. 9–10). Furthermore, it remains unclear why certain of these less-familiar groups failed to give rise to big or anatomically spectacular species in the same way that other Mesozoic marine reptile groups did. Perhaps competition from those other groups prevented this from happening, perhaps the right conditions never became available to them, or perhaps they were simply unable to evolve larger size or become specialized for life in the pelagic realm.

MESOSAURS

Mesosaurs are surprising, since their fossils date to the Permian, and specifically the Early Permian, from 299 to 270 million years ago. Not to be confused with mosasaurs, mesosaurs were long-jawed, long-tailed, anguilliform swimmers. They ranged from 60 cm to 2 m (24 in to 6½ ft) long, and only three kinds are known. These are the relatively large *Mesosaurus*, named in 1864 and known from South Africa, Namibia, Brazil and Uruguay, the smaller *Brazilosaurus* from Brazil and the equally small *Stereosternum*, known from both Brazil and South Africa. However, there are concerns about the distinct nature of these animals and it might be that all *Brazilosaurus* and *Stereosternum* specimens are misidentified individuals of *Mesosaurus*.

When these animals were alive, Africa and South America were connected within the supercontinent Pangaea. The fact that mesosaurs were common to Africa and South America was one of several pieces of evidence used in the 1960s to support the theory of continental drift, proposed by German meteorologist Alfred Wegener, the argument being that such small animals were incapable of crossing the Atlantic. We might challenge that idea today, but convincing evidence for continental drift (involving other fossils and seafloor geology) came from elsewhere anyway.

Several features demonstrate that mesosaurs were aquatic. They had large hands and feet, and some fossils preserve webbing between the digits. The long tail was deep and looked like a sculling organ, and their bones – in particular the ribs and vertebrae – were thick, and almost solid, helping to weigh them down below the surface. However, the anatomy of large, mature specimens suggest that they might have spent time on land, too.

Mesosaurian snouts and jaws are long and crocodile- or gharial-like, and the teeth are long, slender and numerous. Many older reconstructions make their teeth too slender and too numerous, and it's on the basis of such

> On a Middle Triassic reef in Guizhou, southern China, the long-necked *Tanystropheus* – here reconstructed as fully aquatic – forages for fish while a mid-sized *Nothosaurus* pursues small pachypleurosaurs. Pistosaurs (the ancestors of plesiosaurs) swim by at upper right, and the thalattosaur *Anshunsaurus* is visible close to *Tanystropheus*. Bony fishes and ammonites were plentiful in this environment.

The mesosaur *Stereosternum* of Brazil and South Africa. In life, mesosaurs might have resembled small, aquatically adapted crocodiles. They probably had scaly skin, most likely with tiny scales closely packed together. The large mounds in the background are stromatolites: rocky structures built by colonies of micro-organisms.

reconstructions that mesosaurs were once thought to feed on plankton, straining it out of the water through closed jaws. We no longer think this is correct. Instead, their teeth look suited for grabbing shrimp-like crustaceans, the remains of which are found alongside mesosaur fossils. Stomach contents reveal that large mesosaurs also sometimes consumed smaller ones.

Where do mesosaurs fit within the reptile family tree? The mesosaur skull has an archaic look: the palate has a partial covering of small teeth, and the bones between the eye and back of the skull are large, plate-like and appear to lack temporal openings. For these reasons, mesosaurs have mostly been regarded as close to reptile origins. Some experts have argued that mesosaurs belong with other archaic reptiles in the clade Parareptilia, all lineages of which died out during the Permian and Triassic (turtles have also been regarded as parareptiles by some experts). Other experts argue that mesosaurs lack close relatives and are an independent early reptilian branch.

Here we come back to that point about mesosaurs lacking temporal openings. As argued by German palaeontologist Friedrich von Huene in 1941, it turns out that they do have temporal openings, but only the lower one, not the upper one. The presence of only a lower temporal opening is seen elsewhere in synapsids, a group outside of Reptilia altogether, and which includes mammals and their ancestors. So another possibility is that mesosaurs are not reptiles at all. Here, you might remember the point made

Mesosaur skeletons reveal the slender jaws, long, flexible tail, and paddle-like feet typical of this group. Numerous complete, fully articulated mesosaur skeletons are known. This reflects the fact that they often lived and died in highly saline environments where bottom-dwelling scavengers and micro-organisms were rare or absent.

earlier about the ease with which temporal openings can be secondarily closed (see p.28). Maybe we shouldn't rely on the presence or absence of these openings too heavily. The fact that mesosaurs had them could mean they evolved them independently of synapsids and those reptile groups that possess them.

If any of these competing ideas are true, mesosaurs could represent an early, independent invasion of the seas. However, yet another idea exists, and it gives mesosaurs a more 'important' place in reptile history. Superficially, mesosaurs resemble Triassic groups such as thalattosaurs, hupehsuchians and ichthyosaurs. In 2010, marine reptile expert Michael Maisch found that mesosaurs and ichthyosaurs share a list of similarities. Most pertain to the bones in the skull, with the form of the palate and short rear section of the skull being among the most compelling. Could, then, mesosaurs be early members of the superclade? This would extend the origins of this clade back into the Permian – something already suspected to have occurred – and would reduce the number of times in which reptiles adapted to marine life.

Mesosaur biology

Apart from their feeding behaviour, what else do we know about mesosaur biology? A peculiarity of mesosaur tail anatomy concerns what appear to be fracture planes. These are seen in reptiles that shed their tails as a

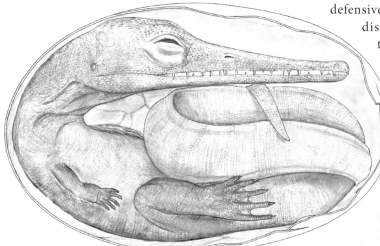

A fossil mesosaur embryo from Uruguay, scientifically described in 2012, is preserved in a curled pose. It was probably contained within a membrane rather than a true eggshell. The baby would have broken out of this membrane seconds after being expelled from the mother's body. The baby was about 15 cm (6 in) long in total.

Claudiosaurus (opposite) from the Permian and Triassic of Madagascar remains an enigma. Direct evidence for its diet is unknown but its short, closely packed, pointed teeth were perhaps used in grabbing crustaceans or small fish. Numerous tiny teeth pepper the bones of its palate.

defensive measure, something that would be disadvantageous to mesosaurs given their reliance on the tail in swimming. Perhaps these fracture planes are a hangover from ancestors that lived terrestrial lives. While this is by no means impossible, it seems odd that such a feature would persist so long after it had any function. Perhaps tail shedding was useful in juveniles, which were sometimes predated upon by adults, or had a role we have yet to appreciate.

Bones from embryonic mesosaurs have been found inside adult mesosaur bodies, revealing how they reproduced. They are too well preserved to be stomach contents, which suggests these creatures were viviparous or, at least, ovoviviparous, where eggs with a reduced or near-absent shell are retained internally for most of their development. Also of note is that juveniles have been found alongside adults, a discovery which has led some researchers to suggest mesosaurs engaged in parental care. An alternative possibility is that mesosaurs were social and gathered in groups, something like Galápagos marine iguanas today.

Finally, while mesosaurs are often described as marine reptiles, some of the places that yield their fossils actually represent gigantic lakes or inland seaways. These might have been open to the sea at times but landlocked at others. Either way, it might be technically wrong to call mesosaurs marine. Geological evidence from South American sediments that yield mesosaurs show that these environments were extremely saline, so the animals that inhabited them must have been highly salt tolerant. Structures preserved on the mesosaurian palate and nasal region look like glands and ducts that would have allowed the excretion of unwanted salt, so their anatomy appears consistent with this.

CLAUDIOSAURUS AND OTHER EARLY DIAPSIDS

As discussed in Chapter 2, it seems that many Mesozoic marine reptile groups belong to the same superclade. However, this is not the case for all of them, and some represent independent marine invasions. Among these is *Claudiosaurus* from the Morondava Basin of Madagascar, an iguana-like reptile about 60 cm (24 in) long. Its skull is small and short-snouted, its hind feet large, and its skeleton is poorly ossified. *Claudiosaurus* was first reported from rocks around 255 million years old and from the Late Permian. However, specimens have since been discovered in Lower Triassic

THE LESSER-KNOWN GROUPS 61

sediments, too, so it survived the end-Permian extinction event. It was described by Canadian palaeontologist Robert Carroll in 1981. Carroll was especially interested in major evolutionary transitions and argued that *Claudiosaurus* was relevant to the origin of sauropterygians, even describing it as an ancestor of this group. This view has been repeated by other authors and has become the 'textbook' view of this animal. There never was, however, any good reason for this view, since *Claudiosaurus* lacks features that link it with sauropterygians. Phylogenetic studies published since Carroll's work find it to be an archaic diapsid, well away from the superclade.

A few other late Permian diapsids from the same approximate part of the family tree, conventionally grouped together as the hovasaurids, also appear to have taken to aquatic life, though apparently in freshwater wetlands, not the sea. Both claudiosaurs and hovasaurids were in the right position in geological history and had the right anatomical features to – theoretically – give rise to sea-going dynasties. But they did not.

TRIASSIC SAUROPTERYGIANS

Among the most familiar of extinct reptiles are the plesiosaurs, an iconic group with a distinctive body plan. Plesiosaurs are closely related to a number of lesser-known groups, all of which were restricted to the Triassic. They are diverse and include the shellfish-crushing placodonts, the long-necked and small pachypleurosaurs, and the mostly long-jawed nothosaurs and pistosaurs. Together, these animals are united in Sauropterygia, meaning winged lizards.

The concept of Sauropterygia arose in 1860, when British anatomist and palaeontologist Richard Owen drew attention to the shared presence in these animals of paddle- or wing-like limbs, teeth implanted in sockets, and large temporal openings. An especially unusual sauropterygian feature concerns the position of the clavicle relative to the scapula. Ordinarily, the clavicle attaches to the outer surface of the scapula. The sauropterygian clavicle is attached to the scapula's inner surface, so it moved 'inside' the bones of the pectoral girdle during evolution.

In fact, numerous features make the sauropterygian pectoral girdle unusual. The clavicles form a bar at the front while the coracoids are enlarged, directed diagonally towards the midline and with extra flanges for muscle attachment. Meanwhile, the scapulae are positioned low on the ribcage and the part that projects upwards (ordinarily forming a large, square plate on the side of the ribcage) is small and slender. Why these changes occurred is not entirely clear. The clavicle bar perhaps prevented distortion or twisting of the pectoral girdle during movement of the forelimbs, while the modified coracoids and scapulae anchored enlarged muscles responsible for pulling the forelimbs forwards, backwards, upwards and downwards.

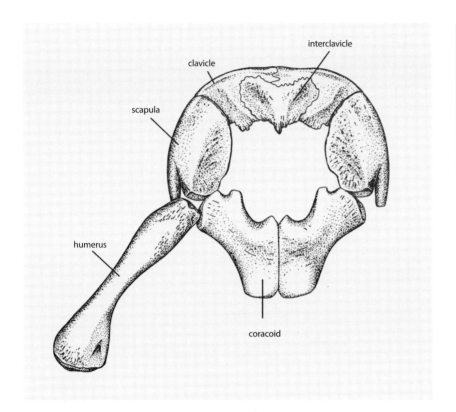

The sauropterygian pectoral girdle – this diagram shows that of a pachypleurosaur, seen from below – is unusual. The thick bar at the front (formed of the scapulae, clavicles and interclavicle) is characteristic. The clavicle is positioned on the inner surface of the scapula, the reverse of the normal situation.

Placodonts

The Triassic saw the appearance and disappearance of many marine reptile groups. Among the most remarkable are the placodonts, named for their flat, rounded teeth (placodont means 'flat toothed'). These animals would have looked something like big, heavily built, swimming lizards or wide, flattened turtles. Most were between 1 and 2 m (3¼ and 6½ ft) long. Placodonts were first recognized in 1833 when French fossil fish expert Louis Agassiz described a partial skull from the German Muschelkalk, notable for the massive teeth on the palate and jaw edges. He named it *Placodus gigas*. As is often the case in placodonts, the fossilized teeth are shiny and black, but would have been white or off-white in the live animal. Agassiz thought that *Placodus* was a fish that could crush shelled prey. This is understandable, since placodont teeth do look like the crushing teeth of certain fish.

The discovery of better remains revealed that *Placodus* was not a fish but a sea-going reptile between 2 and 3 m (6½ and 10 ft) long, equipped with dense, heavy bones, paddle-like limbs and a long tail. Bony nodules ran along its spine. Spoon-shaped incisor-like teeth project from the fronts of its jaws and appear specialized for the plucking of prey such as shelled molluscs, crustaceans and sea urchins. The skull of *Placodus* is thick-boned and looks compatible with the transfer of substantial muscular force to hard objects.

THE SAUROPTERYGIAN FAMILY TREE

Several different ideas exist on how the sauropterygian groups might be related. Placodonts are so odd that their sauropterygian status has sometimes been doubted. However, they possess the key anatomical traits of this group, like that unusual pectoral girdle and a curved upper arm bone. They appear to be one of the first sauropterygian groups to evolve.

The idea that pachypleurosaurs, nothosaurs and pistosaurs might be part of the same group as plesiosaurs has, in contrast, never been controversial, in part because they look like plesiosaur prototypes. Pachypleurosaurs and nothosaurs have sometimes been grouped together, but it seems more likely that nothosaurs are closer to pistosaurs and plesiosaurs. Pistosaurs appear to include the direct ancestors of plesiosaurs.

A few other animals might also belong to Sauropterygia, or at least be closely related to it. Among these is *Atopodentatus* from the Middle Triassic of China, a long-tailed, 3 m (10 ft) animal with flipper-like limbs. The upper jaw of *Atopodentatus* was initially thought to feature two downward-projecting flanges, lined on their inner edges with needle-like teeth. Similar teeth were packed into the lower jaw, which was turned down at its tip. Quite how the animal was using these jaws and its 730 teeth was a mystery. Additional specimens later proved that this interpretation was a mistake. The upper jaw

Atopodentatus, named in 2014, is known from near-complete skeletons. Its skull is its most interesting feature. *Atopodentatus* could probably walk on land. It remains unknown whether it laid eggs or gave birth to live young; both are plausible.

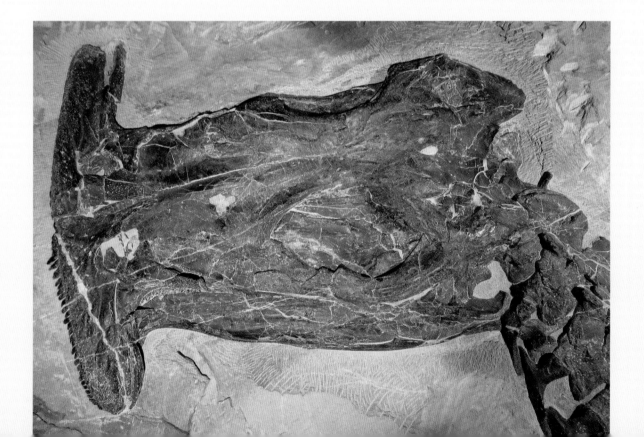

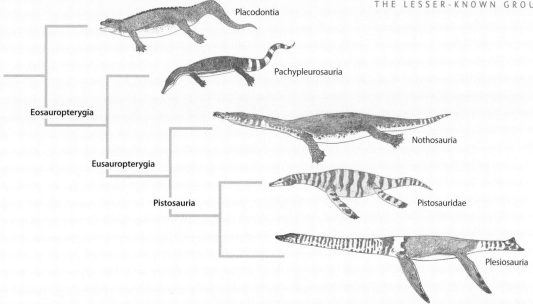

flanges and lower jaw tip did not point downwards but to the side, the result being a hammer-headed look where the teeth were arranged in combs along the mouth's broad leading edge. One suggestion is that *Atopodentatus* was sieving tiny invertebrates from the water. Another is that it scraped at algae on the seafloor and used its tooth combs to retain particles within the mouth.

Thanks to our understanding of the sauropterygian family tree, we can make some generalizations about the biology and lifestyle of these animals. All sauropterygians come from sediments deposited in coastal or fully marine environments, and all have features that link them to an aquatic lifestyle. Sauropterygians appear to have started their aquatic career with a suite of aquatic specializations, these probably including an ability to regulate salt via the use of salt-excreting glands in the head and – probably – the ability to give birth to live young in water.

A consequence of these specializations is that sauropterygians were able to evolve large size and a wide distribution very quickly during the Early Triassic. In fact, they had exploded in diversity and evolved into giants within just five million years of the end-Permian extinction.

> This simplified family tree depicts the relationships between the main sauropterygian groups. Placodonts and pachypleurosaurs are the most 'archaic' of them, and the ones most able to move on land. Nothosaurs and pistosaurids were highly aquatic while plesiosaurs were fully committed to life in water.
>
> In life, *Atopodentatus* probably looked superficially like a nothosaur, its unique 'hammerhead' giving it an utterly unique appearance. The nostrils are located close to the wide front of the mouth while the large eyes are positioned close to the back of the skull.

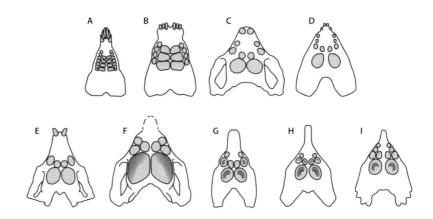

Placodonts were variable in tooth shape and placement. These diagrams show the palate and upper jaw. Peg-like teeth at the front are obvious in *Paraplacodus* (A) and *Placodus* (B). Especially massive, rounded palatal teeth are obvious in others, like *Macroplacus* (F).

Several three-dimensional skeletons of the large placodont *Placodus* are known. The animal's broad cross-sectional shape is obvious, as are the heavily built gastralia on the animal's underside and the row of bony nodules along the midline of the back.

The plucking and crushing teeth of placodonts, their heavy-boned bodies and a shape suited for slow movement close to the seafloor all indicate that they were bottom-feeding shellfish predators of shallow, coastal habitats. An alternative proposal, put forward by palaeontologist Cajus Diedrich, is that placodonts were algae-eaters, something like sauropterygian sea-cows. Diedrich based this idea on the claim that placodonts reduced their teeth over the course of their evolution, and on the suggested presence of algae in places we know placodonts lived. Both alleged pieces of evidence are disputed by all other scientists who've studied these animals, and Diedrich's model is contradicted by evidence better linking placodonts to a diet of shelled invertebrates.

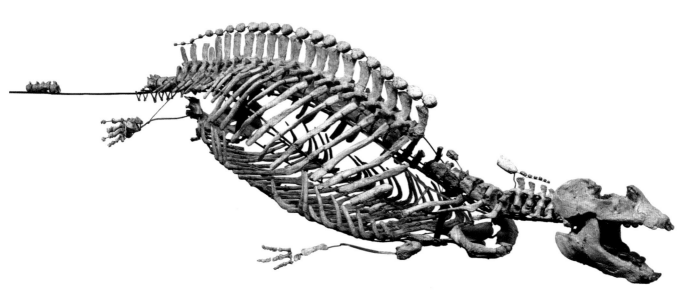

Cyamodontoids, the turtle-like placodonts

New placodonts, including *Cyamodus* and the pointy-snouted *Psephoderma*, became known to science soon after *Placodus*. Unlike *Placodus*, these animals are armoured and belong to the group Cyamodontoidea. *Cyamodus* has a short snout and the back of its skull is wide, equipped with enormous temporal openings, and studded with conical nodules. It was superficially turtle-like, the upper surface of its neck, back and tail covered by a pavement of interlocking armour plates. Triangular plates projected sideways, creating serrated edges to the animal's upper surface. Mobile bands in its armour meant that the hindlimbs, hips and tail had some flexibility relative to the thorax.

A more extreme configuration is present in *Psephoderma* and its kin, the placochelyids. These possessed a flattened, turtle-like shell, with a carapace above and sometimes a plastron below and a side section connecting the two. The placochelyid snout is long, narrow and toothless, as is the lower jaw's tip. These animals might have used their jaws to grab shellfish, perhaps wrenching them from surfaces before crushing them with their rounded teeth. Alternatively, they might have probed into sediment in quest of burrowing crustaceans and other prey. Their skull bones are less firmly connected than those of other placodonts, perhaps meaning that they consumed softer – or at least less hard – prey than most of the others. Placochelyids were rarely more than 2 m (6½ ft) long but isolated teeth show that some reached 3 m (10 ft).

Some experts have argued that placochelyids might be imagined as 'reptilian rays', and that they lived a life similar to that of living eagle rays. Some similarities between these groups do exist – rays are flattened, wide-bodied predators of shellfish – but the differences are prominent and obvious. Rays are not protected by a shell, they move with giant fins and they don't visit the surface to breathe. Placodonts weren't exactly like any living animals, but the closest similarity might be with bottom-feeding turtles.

A few other animals are close relatives of *Placodus* and the armoured cyamodontoids. Among them is *Paraplacodus* from Monte San Giorgio and perhaps the Muschelkalk of Germany and elsewhere. Long, pointed teeth project from the fronts of the jaws, and the cheek region is deeply embayed. A tiny skull from the Muschelkalk of the Netherlands – just 2 cm (¾ in) long – represents an especially early member of the placodont lineage. This is *Palatodonta*, and the small size of the one known specimen is partly due to its juvenile status. The skull slopes from the tall forehead to the short, shallow snout, and the cheek region is deeply embayed like that of *Paraplacodus*. Palatal teeth are present but are pointed and different from those typical of placodonts. They are arranged in comb-like fashion, suggesting that *Palatodonta* sifted through the seafloor mud in quest of small invertebrates.

The fact these most archaic members of the placodont group (the whole of which has been termed Placodontiformes) are European suggests that the

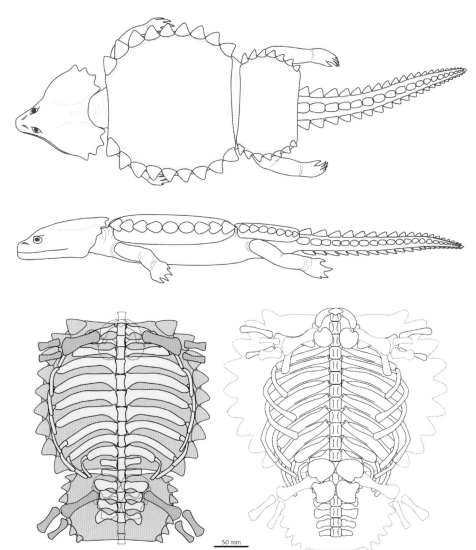

Cyamodus was a remarkable reptile. Several species occurred across Europe during the Middle Triassic and in China during the Late Triassic. The tremendous breadth of the back of its skull is notable, as is its flattened, spiky-edged body armour.

group originated in the Western Tethys Ocean. On that note, placodonts were known almost exclusively from Europe until the late twentieth century, the majority being from Germany, Switzerland and Italy, though some are from Spain, England, Hungary, Romania, Turkey, Israel and countries in northern Africa. Were placodonts absent from the Eastern Tethys Ocean, or is a poor fossil record to blame?

This question was answered in 2000 with the discovery of China's first placodont, the cyamodontoid *Sinocyamodus*. Other Chinese placodonts have since been reported, including additional cyamodontoids as well as a species of *Placodus*. These fossils suggest that all placodont lineages were present in both the Western and Eastern Tethys. It is currently assumed that placodonts originated in the west and moved east, but this is by no means clear and more fossils are needed to better understand their history.

The best way to end the placodont story is with the weirdest placodont of all, and indeed one of the strangest of sauropterygians. This is *Henodus* from the Late Triassic of southern Germany. *Henodus* has a flattened, short-faced skull where the nostrils and eyes are close to the broad mouth, an extremely wide, flattened carapace, paddle-like limbs and a short tail. It did not possess big, crushing teeth, but instead had grooves along the edges of its jaws and a row of tiny, square denticles arranged on a flange along the leading edge of its upper jaw. The grooves appear to have contained vertically aligned fibres that recall whale baleen.

It seems that *Henodus* was adapted for scraping or biting soft material with the denticles, and then using the baleen-like material in sieving or filtering. Perhaps it dug for small invertebrates in mud, or maybe it scraped at algae and filtered the particles from the water. *Henodus* inhabited a giant, semi-enclosed, brackish lagoon rather than the sea, so it wasn't a marine reptile in the proper sense. A closely related, similar animal – *Parahenodus* – was described from Spain in 2018.

> *Henodus* is one of the most turtle-like of placodonts. Several features of the cyamodontoid carapace look hydrodynamic, and it might be that some of these animals were speedy swimmers, as implied here. However, a slow and lazy life is equally plausible, especially for lagoon-dwelling *Henodus*.

Pachypleurosaurs

Sauropterygians are mostly weird, surprising or spectacular. It is arguable, however, whether this applies to pachypleurosaurs, a Middle and Late Triassic group associated with Monte San Giorgio and Germany, but known since the 1950s to have a presence in China, too. Pachypleurosaurs were moderately long-necked and perhaps resembled large, stretched, aquatic lizards – the biggest reached 1.2 m (4 ft). Their name means thick-ribbed lizard and refers to their thick, dense-boned, tightly packed ribs.

It has often been thought that pachypleurosaurs were anatomically unmodified relative to the earliest sauropterygians. The quadrate bone at the back of the skull shows that they had a large eardrum and air-filled middle ear, and were thus listening to air-borne sounds and presumably spending

This pachypleurosaur embryo – preserved in the 'curled' posture typical of new-born reptiles – was reported from the Middle Triassic of Monte San Giorgio in 1988. It is 5 cm (2 in) long in total with a proportionally enormous skull.

time on land. The small, pointed teeth and delicate jaws look suited for grabbing small fish and invertebrates. So far so normal.

In some respects, however, pachypleurosaurs are unusual. Their pectoral girdle has the specialized sauropterygian form. The tail is usually only slightly longer than the body and lacks features associated with a specialized role in swimming. The limbs are not unusual, but the upper arm bone is surprisingly robust and often curved. The neck (sometimes formed of as many as 26 vertebrae) was flexible and could bend far to the side. In the skull, the snout is short and rounded in some species but sharper and longer in others, the eyes are large, and the throat bones (or hyoids) are well developed. These hyoids – combined with the relatively short jaws and evidence for rapid jaw-opening abilities – suggest that pachypleurosaurs used suction to capture prey. Together, these features indicate that pachypleurosaurs were generating some of their propulsion during swimming by undulating from side to side, but were mostly using rowing or paddling motions. They were likely slow, manoeuvrable, and in the habit of foraging close to the seafloor or among reefs.

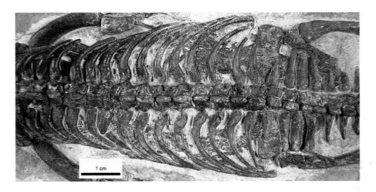

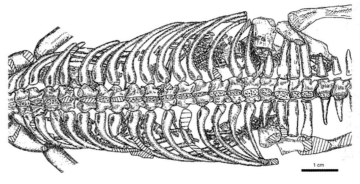

Several pachypleurosaur species are known from specimens collected from the same location. These fall into two groups. One consists of large individuals with thick, curved arm bones and another involves smaller individuals with slimmer arms. These groups presumably represent the sexes, but which is which? Two pregnant specimens of the Chinese *Keichousaurus* were reported in 2004, and both belong to the small, slim-armed group, so this issue is now resolved. These mothers are preserved with four and six embryos each, all of which lacked shells. They therefore demonstrate viviparity in this animal. An isolated embryo of the European *Neusticosaurus* – discovered in sediments deposited far from land and also lacking eggshell – provides additional support for pachypleurosaur viviparity.

Demonstrating viviparity in pachypleurosaurs is significant, since it suggests that viviparity evolved early in sauropterygian history, apparently at a time when members of the group were small and with terrestrial capabilities. However, it is not clear that all pachypleurosaurs were viviparous. Evidence from growth style and body size suggests that some grew at rates more consistent with hatching from eggs.

Several pachypleurosaur specimens preserve embryos inside their bodies. In this photo and interpretative diagram, the mother's head is to the left and the bases of her limbs are visible at the edges. The embryos are present on both sides of her body. The thick, heavy ribs that give this group its name are obvious.

Nothosaurs

After plesiosaurs, the best-known sauropterygians are the nothosaurs, a mostly Tethyan group that tend to be long-bodied, long-necked and with a long, shallow skull equipped with interlocking fangs. The smallest were around 1 m (3¼ ft) long while the largest were giants of around 6 m (19¾ ft) where the skull was up to 75 cm (29½ in). Nothosaurs have often been imagined as evolutionary intermediates between pachypleurosaurs and plesiosaurs, and there has been a tendency to think of them as prototypes of the plesiosaur condition. This is not accurate. They were a distinct group all of their own, with their own specializations.

The term nothosaur was used loosely in the past, and applied to pachypleurosaurs, nothosaurs proper, and pistosaurs too. Today, the term nothosaur is restricted to the clade that includes the species of *Nothosaurus* and *Lariosaurus*, as well as the relatively short-snouted *Simosaurus* and some related forms. A complication is that our current use of the names *Nothosaurus* and *Lariosaurus* do not correspond to discrete groups of related species, but instead to animals widely spread across the nothosaur family tree, not all of which are close relatives. It is expected that a revised naming system will be published soon.

Nothosaurs were originally described from the Muschelkalk and later from the Middle Triassic of Italy, Switzerland, Poland and Austria. During

> The shapes of nothosaur limbs and bodies suggest that they foraged far and wide. They were not static or stationary hunters that relied on ambush. The internal anatomy of their bones indicates that they were warm-blooded and probably able to maintain a high level of activity.

the twentieth century, discoveries made in Israel, Romania, Spain, the Netherlands and elsewhere demonstrated their presence throughout the Western Tethys Ocean. Their absence from the Eastern Tethys Ocean was always suspicious. Was this due to absence or simply a lack of discovery? In 1965, a nothosaur was reported from the Middle Triassic of China. This proved to be a *Nothosaurus*, and it heralded the discovery of numerous Chinese finds. Today we know of many Eastern Tethyan nothosaurs, large and small, and it seems that the coastal waters of China were important in their history.

Did nothosaurs occur outside the Tethys Ocean? Nothosaur bones are known from the Lower Triassic of British Columbia, Canada, an area that was on the eastern edge of Panthalassa at the time. This could mean that nothosaurs occurred right around the northern margins of Pangaea. Alternatively, it might be that a short-lived seaway allowed these animals to move between the Western Tethys Ocean and Panthalassa.

Nothosaurs have sometimes been thought of as reptilian seals: amphibious, lizard-shaped predators equipped with a crocodile-like head. However, thickened, dense-boned vertebrae, ribs and limb girdles suggest that they hunted on or near the seafloor. Furthermore, the skull of *Nothosaurus* – the best known nothosaur – is unique. It is long, shallow, and with a narrowing that separates the tip from the rest of the face. Interlocking fangs line the jaws, those at the front being far longer than those at the back, and bony bars and openings at the skull's rear reveal evidence for muscles that opened and closed the jaws extremely quickly. The palate is continuous right to the rear of the skull, except for openings mid-way along and two side openings at the back. Maybe this extensive palate helped dissipate stress conducted during biting.

Nothosaurus probably caught prey by making swift sideways snapping actions. It might be that prey were impaled on the teeth, or that the teeth formed a cage. Stomach contents show that fish were eaten, but so were smaller marine reptiles including pachypleurosaurs and placodonts. The giant nothosaurs known from central Europe, Israel and China were the biggest predators in their environments, and their size alone shows that they were almost certainly preying on marine reptiles around 3 m (10 ft) long.

Simosaurus of the Muschelkalk, and also the Upper Triassic of Italy, is different. Its jaws are shorter than those of *Nothosaurus*, its skull is broader and more robust, and its teeth have blunt tips. Presumably, it preyed on shelled or armoured prey, and its jaws are wide enough and its musculature powerful enough to suggest it might have used suction to capture agile prey, such as ammonites or fishes. Given that it reached 3 to 4 m (10 to 13 ft), *Simosaurus* could also have grabbed larger animals such as other marine reptiles. *Paludidraco* from the Late Triassic of Spain, a close relative,

The nothosaur skull is typically long and narrow, with elongate openings at the back for large jaw-closing muscles. The eye sockets are directed upwards somewhat, but still allowed a wide field of sideways vision. The snout's tip is separated from the rest of the face by a distinct narrowing.

Like all nothosaurs, *Simosaurus* looks built for slow swimming in shallow marine environments. First described from the Muschelkalk of France in 1842, it is best known for fossils found in Germany. This skeleton is on display in Stuttgart, Germany.

possessed dense-boned, heavy ribs with broadened ends, a delicate lower jaw and numerous small, closely spaced, curved teeth. Its heavy bones show that it mostly foraged at the seafloor where it perhaps sieved small animals from the sediment, or fed on algae.

The nothosaur pectoral girdle is of the typical sauropterygian sort, built to allow powerful backstrokes of the robust forelimbs. Because nothosaurs had flexible elbows, wrists, knees and ankles, it has often been thought they were amphibious, and able to rest on rocks and walk on beaches. However, their hip bones are small and would have functioned poorly in transferring weight, and their limbs are flattened and built for rowing more than walking. *Lariosaurus* has more finger bones than usual – four or five, as opposed to the typical three. For this reason, it has been suggested it had flipper-like forelimbs. Hundreds of Middle Triassic tracks from China, made by nothosaurs swimming just above the seafloor, confirm this view. Flipper-like limbs were not unique to *Lariosaurus* but present in nothosaur species large and small.

We saw earlier that pachypleurosaurs were mostly viviparous, so perhaps nothosaurs were too. Possible evidence for this comes from the discovery of four *Lariosaurus* embryos preserved without any eggshell. In a 2019 study, Eva Griebeler and Nicole Klein looked at growth rates (via analysis of bone sections) and size in as many *Nothosaurus* specimens as possible. In living reptiles, viviparous animals are born larger and reach maturity later than those that hatch from eggs. The *Nothosaurus* data indicates that both viviparity and egg-laying were present in the group, as is also the case in pachypleurosaurs. We might imagine that some nothosaurs, then, were capable of movement on land to visit beaches to dig nests, whereas others were completely aquatic.

Pistosaurs

Pachypleurosaurs and nothosaurs share traits with plesiosaurs. But both groups are very different from plesiosaurs, either because they are small and generalized, or because they have their own specializations. We do, however, know of Middle and Late Triassic animals that bridge the gap between these groups and plesiosaurs, and reveal how the plesiosaurian body plan emerged. These are the pistosaurs. They take their name from *Pistosaurus*, described from a long-snouted skull reported from the German Muschelkalk in 1839, by German palaeontologist Hermann von Meyer. Uncertainty surrounds the origin of the name *Pistosaurus*, but researcher Ben Creisler has noted that von Meyer might have been referring to the bottle-like shape of the skull, since one translation of *Pistosaurus* is 'drinkable lizard'!

Pistosaurus has larger, longer coracoids than nothosaurs, simpler-shaped limb bones, and a shorter tail. These features are also evident in the American pistosaur *Augustasaurus* as well as in *Bobosaurus* from Italy. *Bobosaurus* has surprisingly tall neural spines (meaning it had an especially deep neck, body and tail) and long flippers and must have looked very plesiosaur-like. China is currently the most significant location for pistosaurs. *Yunguisaurus* has

The near-complete skeleton of the Chinese pistosaur *Wangosaurus* reveals the long neck and short tail of this animal. Some of its toes possess extra bones relative to the 'normal' sauropterygian condition, one of many features that presages those of plesiosaurs.

long, gently curved fingers and a broadened ulna, both of which suggest that its forelimb bones were united in a slender flipper. Its massive eyes and long, slim teeth give its skull a plesiosaurian look. Another Chinese pistosaur – *Wangosaurus*, named in 2015 – has a long neck, with 33 vertebrae, and a skull where everything behind the eyes is narrow and stretched. None of these animals were large, mostly 2 to 3 m (6½ to 10 ft) long. Their slender snouts and slim, pointed teeth suggest reliance on fish, squid and crustaceans.

Pistosaurs are not a discrete group but are 'near-plesiosaurs' that includes plesiosaur ancestors. Their strongly muscled pectoral girdles and long, streamlined limbs show that they swam with rowing or paddling motions, and were adapted for the open ocean more than other Triassic sauropterygian groups.

Plesiosaurs evolved from pistosaurs – probably from a *Bobosaurus*-like animal – around 240 million years ago during the Middle or Late Triassic, and their key novelties included broader, more wing-shaped limbs and even bigger coracoids. We usually think of pistosaurs as one of the many sauropterygian groups that failed to survive the extinction events of the Triassic. In fact, they did survive, albeit as the modified group we call the plesiosaurs.

TANYSTROPHEUS AND KIN

Among the most remarkable of Triassic marine reptiles – and most remarkable of reptiles ever – is the large, long-necked *Tanystropheus*. Named by Hermann von Meyer in 1855 for fossils from the German Muschelkalk, *Tanystropheusi* is part of an archosauromorph group termed the protorosaurs or prolacertiforms. These were mostly terrestrial, quadrupedal omnivores and predators, some of which were close to the ancestry of archosaurs. Protorosaurs are probably not a clade but an assemblage of archaic archosauromoph lineages.

Several protorosaurs look like close relatives of *Tanystropheus* and are united with it in the clade Tanystropheidae. They are typically 1 m (3¼ ft) long or less and possess long, slender hindlimbs, a slim, flexible neck, a shallow snout, large eyes and numerous recurved teeth. *Tanystropheus* is

The idea that *Tanystropheus* might be aquatic has been challenged on a few occasions, partly because its proportions look better suited for a life that involved regular walking. This reconstruction also emphasises the incredible length of the neck.

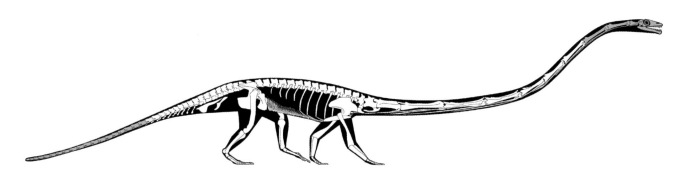

gigantic relative to these others, growing to around 6 m (19¾ ft), but it is its neck that usually forms the focus of discussion. This is formed of 11 long, tubular vertebrae, plus two shorter vertebrae at the neck's base, the result being a neck similar in length to the body and tail put together. The fact that this neck is mostly formed of long, straight vertebrae means that its flexibility was limited, though vertical movement was possible at its base and at the joint between the head and neck. The neck was as lightweight as possible, yet might still have accounted for a fifth of the animal's weight.

Tanystropheus was not just an oddity, but a successful animal, species of which existed from the Early Triassic until the Late Triassic Carnian age, a duration of around 20 million years. After its discovery in Germany, specimens were found in Poland, Italy, Switzerland, Spain, France, Romania, Israel, Saudi Arabia, China and Canada. This shows that *Tanystropheus* lived along the western and eastern margins of the Tethys Ocean, including on islands several kilometres or miles from land, and in estuaries and rivers. Its presence in Saudi Arabia shows that it occurred along the eastern margin of Gondwana, perhaps as far south as India.

Tanystropheus also occupied a range of sizes. Small specimens thought to be juveniles, around 1.5 m (5 ft) long, have a bone structure that shows

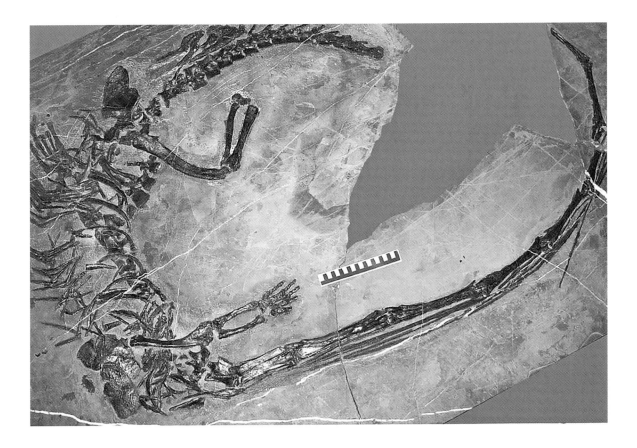

This Chinese fossil, described from Guizhou in 2014, is one of several demonstrating the presence of *Tanystropheus* in the Eastern Tethys. The skull is missing but the extraordinary neck emerges towards the right before curving upwards. This specimen is a 'mid-sized' *Tanystropheus* with a total length of around 3 m (10 ft).

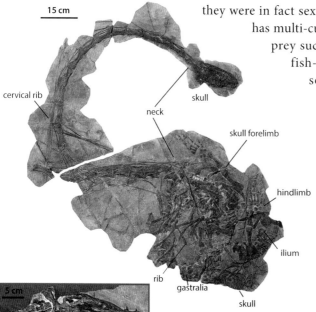

This articulated *Dinocephalosaurus* skeleton, described in 2004, has a neck 1.7 m (5½ ft) long but a body less than 1 m (3 ft) long. Its skull is beautifully preserved and reveals a flattened snout with a squared-off tip.

they were in fact sexually mature. A small species from Switzerland has multi-cusped cheek teeth, so was presumably exploiting prey such as crustaceans, different to its larger, mostly fish-eating, cousins. Studies have also shown that some fossils once included in *Tanystropheus* should be treated as separate animals, including *Amotosaurus* from the Middle Triassic of Germany and *Sclerostropheus* from the Late Triassic of Italy.

The association *Tanystropheus* has with sediments deposited at sea means that experts have generally assumed it was marine. This is backed up by cephalopod and fish remains found in the stomach and gut regions of some specimens, and also by its skull. The long fangs at the front interlock in a manner typical of animals that grab prey in water, the nostrils are on the snout's upper surface, and the skull is streamlined in a manner suited for rapid sideways movement underwater.

In some respects, *Tanystropheus* doesn't look marine. Its tail is broad rather than deep and is not at all paddle-like, and its limbs look suited for walking more than swimming. A popular suggestion is that it was an animal of the shoreline, and that it sat on shores and coastal rocks, using its neck like a fishing pole. Good walking abilities would by no means be inconsistent with a seaside ecology or a head adapted for the grabbing of aquatic prey.

Dinocephalosaurus from the Middle Triassic of China is similar to *Tanystropheus*, but also very different. It was 3.5 m (11½ ft) long and also has an immensely long neck. But while the *Tanystropheus* neck is formed of 11 long and two short vertebrae, that of *Dinocephalosaurus* is formed of 27 short vertebrae. Its long, paddle-like hands and feet and poorly ossified wrists and ankles show that it was aquatic, as does its preservation in marine sediments known for abundant fishes, ichthyosaurs and other marine reptiles.

Based on the anatomy of its skull and neck, *Dinocephalosaurus* probably used its neck to approach prey, most likely fish, before using suction to draw it into its mouth. One specimen is preserved with an embryo inside its body, this demonstrating viviparity. Here is proof that archosauromorphs could evolve this habit, something relevant to thoughts on another marine archosauromorph group: the thalattosuchians (see Chapter 7). Perhaps aquatic protorosaurs evolved viviparity after they became aquatic. We assume, but cannot yet confirm, that *Tanystropheus* was viviparous too. Or perhaps viviparity was widespread in these animals and eased their transition to aquatic life. Indeed, *Dinocephalosaurus* belongs to a different branch of the family tree from *Tanystropheus*, which means that radically long necks evolved at least twice within protorosaurs. The *Dinocephalosaurus* group – termed Dinocephalosauridae – includes another marine, long-necked form from China, the slender-snouted *Fuyuansaurus*.

PHYTOSAURS AT SEA

We know of another Triassic archosauromorph group that made at least two marine excursions during its evolutionary history - namely, the phytosaurs. Phytosaurs are restricted to the Triassic and lived virtually worldwide, so were a long-term, persistent presence in Triassic wetlands. The best known phytosaurs were superficially crocodile-like and are often discussed as textbook examples of convergent evolution. They weren't closely related to crocodiles or even crocodylomorphs yet evolved a similar shape and lifestyle. A key difference between phytosaurs and crocodylomorphs is that phytosaur nostrils are located on a bony mound close to the eyes, rather than at the snout's tip.

A block of rock containing the skull and lower jaw of the marine phytosaur *Mystriosuchus*, as discovered in the Upper Triassic of Austria in 1980. The two halves of the jaw are joined for much of their length, the result being a Y-like shape. Jaws of this sort evolved numerous times in aquatic vertebrates.

In the *Mystriosuchus* skull, the snout and remainder of the skull are markedly different in height. The eye socket and nostril are high up on the skull. The oval opening close to the base of the upper jaw is the antorbital fenestra. In life, this was occupied by an air-filled structure connected to the respiratory system.

Mystriosuchus inhabited shallow marine seas, perhaps often foraging tens of kilometres from the nearest shoreline. The remains of several individuals have been discovered together, this hinting at mass death due to some kind of local disaster.

Virtually all phytosaur fossils come from sediments laid down in rivers, lakes and swamps, but at least two members of the group appear to have been marine. The first is *Diandongosuchus* from China, an especially archaic phytosaur, different in shape and proportions from typical members of the group. The second is *Mystriosuchus* from the Late Triassic of Italy and Austria, a slender-snouted 'typical' phytosaur that evolved late in the history of the group. It was about 4 m (13 ft) long, had a gently upturned rostrum and would have looked superficially like a gharial. Numerous fossils of this animal have been discovered in sediments laid down in warm, shallow lagoons, enough to indicate that it was a resident of this sort of environment, not merely a visitor.

On that note, it is surprising that phytosaurs only took to marine life twice, since they look ideally suited for it. Maybe the existence during the Triassic of ichthyosaurs, sauropterygians and other groups prevented their movement into the marine realm.

MARINE PTEROSAURS AND BIRDS

Two geologically long-lived marine archosaur groups are ignored in this book. Firstly, pterosaurs: a Mesozoic group related to dinosaurs, many species of which were similar to gulls, albatrosses or frigatebirds in their adaptation to marine life. Numerous Jurassic pterosaurs have long jaws and needle-like teeth and come from the same Tethyan sediments as ichthyosaurs, plesiosaurs and thalattosuchians.

Cretaceous pterosaurs reached much greater sizes and include species with long skulls where fang-like teeth are restricted to the jaw tips, as well as the toothless, fabulously crested pteranodontids and nyctosaurids. Evidence that these animals were marine comes from the sediments in which their fossils are found, their stomach contents (they were eating marine fish and squid), and their association with plesiosaurs and other marine groups.

Secondly, seabirds. In non-scientific language, birds are considered distinct from reptiles. But in scientific language – where organisms are grouped together on the basis of ancestry – birds are a subset of theropod dinosaurs and are therefore archosaurian reptiles. Seabirds include the flightless penguins, the fast-flying, diving auks and diving petrels, the piratical skuas, the adaptable, mostly coastal gulls, and those supreme aerialists the terns, frigatebirds, petrels and albatrosses.

The majority of birds we call seabirds belong to the group Aequorlitornithes, and numerous other birds associated with water belong to this group, too, including flamingos, waders, herons and storks. However, marine habits evolved in other groups both modern and extinct, including in toothed groups of the Mesozoic (like the flightless, foot-propelled hesperornithines of the Cretaceous), in ducks, in the giant, extinct bony-toothed birds, and in kingfishers. A huge amount could be said about marine birds and what we know about their evolutionary history, but this book isn't the place for it.

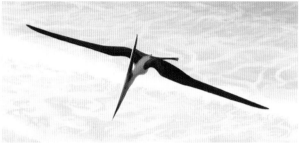

Birds and pterosaurs have both given rise to long-winged, ocean-going specialists. Among the largest of these are albatrosses (top) and pteranodontids (middle). Unlike pterosaurs, birds have repeatedly given up on flight and flightless seabirds like the hesperornithines (bottom) have been a constant presence since the Cretaceous.

HUPEHSUCHIANS

Until just a few years ago, hardly anyone beyond a handful of specialist researchers had ever heard of the Triassic hupehsuchians, a group so far restricted to China. This has changed with the twenty-first-century discovery of intriguing new species, and with suggestions that they had a role in ichthyosaur origins. Hupehsuchians were limited to the Early Triassic and appear to have been limited to the Eastern Tethys Ocean. They were superficially similar to a few other Triassic marine reptile groups. They

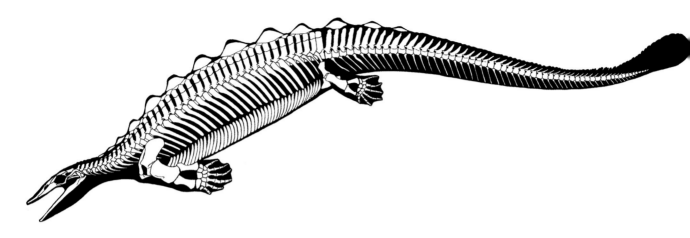

Eretmorhipis, named in 2015, is among the strangest marine reptiles of all. Its name means 'oar fan' and refers to the broad shape of its limbs. Several specimens are known, meaning that its bony anatomy is comparatively well understood.

were mostly around 1 m (3¼ ft) long and had a long, laterally compressed tail, long body, and paddle-like limbs with spreading digits connected by webbing. At least one member of the group possessed an unusual number of digits, with seven on each hand and six on each foot. Extra digits evolved elsewhere in Mesozoic marine reptiles (most notably in parvipelvians), probably to make the limbs broader.

Three other features make hupehsuchians unusual. The first is that they were toothless, most species possessing long, flattened jaws. The second is that they were extremely 'bone heavy': the ribs are broad and in contact all along the body, and the gastralia are, similarly, in close contact. Hupehsuchian vertebrae have a 'blocky' look and the neural spines that project upwards from them are robust and closely spaced. The result is a stiff, tubular body. The third unusual feature is a row of horn-covered bony plates (known technically as osteoderms) along the back's midline. Osteoderms are not typical for Mesozoic marine reptiles and are more associated with archosaurs, so an early idea on hupehsuchians is that they might be marine archosaurs.

What might these features tell us about hupehsuchian lifestyle? Their paddles look suited for slow manoeuvring and are suggestive of life in cluttered environments such as weed-choked gullies or among masses of floating invertebrates. Rafts of gigantic, stalked animals called crinoids existed in Triassic seas and were frequented by marine reptiles, fishes and other animals. Perhaps hupehsuchian paddles were used to clamber across uneven seafloors, or among plants or crinoid rafts.

Hupehsuchians have been known since 1959 when *Nanchangosaurus* was named, but *Hupehsuchus* was the first member of the group to receive wide attention, named in 1972 but made the subject of broader interest in 1991. Others have been named more recently, including *Eohupehsuchus* and the remarkable *Eretmorhipis*. *Eretmorhipis* has the features characteristic of this group, but combines them with a small head, short neck, especially broad limbs and unusually large osteoderms. It must have looked something like a mythical Chinese dragon, albeit an aquatic one only 90 cm (35 in) long. Its osteoderms give it a serrated outline and are reminiscent of the scutes, or external horny plates, present in modern sturgeon. Sturgeon scutes have a mostly defensive role, so maybe this was the case in *Eretmorhipis* too. However, sturgeon scutes differ between species, change shape as the animal grows and might be used in sending information on age and mating condition. We can only speculate whether this was also true for *Eretmorhipis*.

A major question about hupehsuchians is where they fit within the reptile family tree, and indeed this was framed as *the* main point of interest about them in a study published in 1991. Vertebrate palaeontologists Robert Carroll and Zhi-Ming Dong argued that hupehsuchians were an enigma. While possessing features similar to those of thalattosaurs, ichthyosaurs and sauropterygians, they lack the key traits of those groups and couldn't, they suggested, be considered allied to any of them. Carroll and Dong's conclusion was that working out the affinities of Mesozoic marine reptiles might be doomed to failure, the spectre of evolutionary convergence obscuring any chance of finding clear connections. Further study has shown that

Eretmorhipis perhaps foraged in seafloor sediment, though its exact diet and feeding strategy remain unknown. Its closely-packed ribs and gastralia and scutes suggest that it was relatively heavy, and hence perhaps adapted to remain at depth.

hupehsuchians do, in fact, possess features that allow us to determine their evolutionary affinities. Specifically, they share forelimb details and unusual features of the vertebrae with ichthyosaurs. When included with other Mesozoic marine reptile groups in evolutionary analyses, hupehsuchians are members of the superclade (see Chapter 2), close to ichthyosaur ancestry.

THALATTOSAURS

Another obscure Triassic group are the thalattosaurs, known formally as Thalattosauria or Thalattosauriformes. Thalattosaurs became known in 1904 when American palaeontologist John Merriam (better known for his ichthyosaur studies, see pp. 105–107) described a new marine reptile from the Late Triassic of California. Initial finds made Merriam think he was dealing with ichthyosaurs, specifically mixosaurs. But more complete remains discovered by patron of the research Annie Alexander, revealed the unusual nature of this group. Merriam named this animal *Thalattosaurus* (meaning 'ocean lizard'). It was about 2 m (6½ ft) long, long tailed, and had vertebrae indicating lateral flexibility and an anguilliform swimming style. A second, smaller species of *Thalattosaurus* has since been discovered in British Columbia, an area that has proved a hotspot for thalattosaur discoveries. This was a highly active volcanic region when thalattosaurs were alive, but what relation this volcanicity had to thalattosaur lifestyle and biology is unknown.

Thalattosaurus's skull is streamlined, large-eyed and narrow across the snout. More unusual is the long, downcurved rostrum, which possesses a

The beautiful and complete skeleton of the Italian thalattosaur *Askeptosaurus* reveals the long, flexible tail, narrow snout and numerous pointed teeth of this animal. In contrast to many other thalattosaurs, it is anatomically unspecialized.

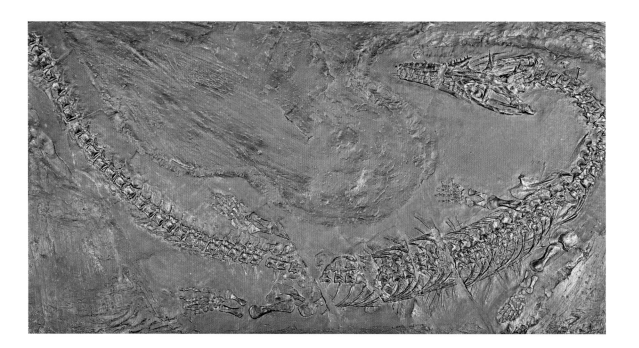

texture suggestive of a beak-like covering and bony pseudoteeth near its tip. Conical teeth are present near the lower jaw's tip while flat, rounded teeth are located further back and across part of the palate. These features suggest that *Thalattosaurus* extracted prey from burrows or crevices before crushing them in the mouth. Perhaps the covering on the rostrum allowed the animal to forage among rocks or reach into rocky crevices without damaging its face.

Additional thalattosaurs were described from Monte San Giorgio during the 1930s. These are *Hescheleria* and *Clarazia*, both of which are shorter-snouted than *Thalattosaurus*. *Hescheleria* has a strongly downcurved snout where a long gap separates the pointed teeth at the tip from rounded teeth further back. It also has a midline bony protrusion at the front of its lower jaw. It is tempting to think that the snout's shape is due to distortion, but the good condition of the bones shows otherwise. Furthermore, other thalattosaurs (namely *Nectosaurus* and *Paralonectes*) with similar snouts are known from California and British Columbia.

The 1930s also saw the description of *Askeptosaurus* from Italy, a large thalattosaur with long, shallow jaws and numerous similarly shaped, sharp teeth. *Endennasaurus*, also Italian, has a pointed, toothless rostrum. Hardly any new thalattosaurs were named between the 1930s and the twenty-first century, but the 2000s saw the naming of *Xinpusaurus* and *Concavospina* from China. The snout in these animals is pointed and toothless, while conical teeth are located further back, those under the nostril pointing forwards. In *Miodentosaurus*, also from China, the snout is short, and conical teeth are present at the jaw tips.

Clearly, thalattosaurs were practising different lifestyles, an idea backed up by variation in size. Most were 1 to 2 m (3¼ to 6½ ft) long but some approached 5 m (16½ ft). *Askeptosaurus* and its kin were built for a diet of other reptiles, fish and crustaceans while *Endennasaurus* was perhaps eating softer prey, perhaps squid. Thalattosaurs with rounded or blunt teeth were likely grabbing shellfish.

The long thalattosaurian tail served as the main swimming organ. Were thalattosaurus fully aquatic? The clawed limbs and well-ossified wrists and ankles of some, like *Endennasaurus*, suggest they were able to walk on land.

Thalattosaurs with down-curved snouts and jaws – like this *Hescheleria* (left) – perhaps used these to pry shellfish from rocky surfaces, or to reach into cavities. A long, flexible body and tail is typical of the group.

Thalattosaurs are diverse in skull and tooth shape. From top to bottom, this selection shows an unnamed Chinese form, *Nectosaurus* and *Hescheleria*. All have down-turned snouts and specialized teeth. The large eye socket, 'open' structure to the rear part of the skull and heavily built lower jaw are also obvious.

However, the flattened limbs and tall tail of *Gunakadeit* from Alaska indicate it was poor at moving on land, perhaps incapable of it. Maybe thalattosaurs were variable in how aquatic they were, and maybe those with 'terrestrial-looking' features were not typical of the group.

To date, thalattosaurs are known from Europe, China and western North America. It is difficult to know what this distribution means, and their absence from places such as Spitsbergen is surprising. They were clearly present throughout the northern Tethys Ocean and eastern Panthalassa, but explaining how they came to be present in both regions is not possible with the fossils we have. Were they present right around northern Pangaea? Did they use a marine channel to move between the Panthalassa and Tethys Oceans? Did they migrate across Panthalassa to get into the Tethys Ocean, or vice versa? Or were they present along all the coasts of the Triassic world? Despite their variation in habitat choice and lifestyle, they failed to persist beyond the Triassic, their extinction around 210 million years ago likely being due to changing sea levels and the impact this had on coastal communities.

SAUROSPHARGIDS, HELVETICOSAURS AND KIN

Another group scarcely known outside the fossil reptile research community is Saurosphargidae of the European and Asian Middle Triassic. Saurosphargids take their name from *Saurosphargis* of Poland, Germany and the Netherlands, the first member of the group to be recognized. The original Polish specimen consisted of part of the spine, some ribs and armour plates. They showed that *Saurosphargis* had a broad, flattened body where the ribs were in close contact and protected on their upper surfaces by rounded osteoderms. This configuration is similar to that of leatherback turtles and explains why the name chosen for this animal combines an old scientific name for the leatherback (*Sphargis*) with the Greek for lizard. Friedrich von Huene described this animal in 1936, and implied that *Saurosphargis* was an evolutionary intermediate between lizard-like reptiles and turtles. Unfortunately, the fossil was destroyed during a wartime siege in 1945, the result being that it was then mostly forgotten about.

The enigma of *Saurosphargis* was resolved in 2011 when Chinese palaeontologist Li Chun and colleagues described similar animals from China. These include articulated skeletons with complete skulls and intact armour. They confirm the presence of a turtle-like look to the broad, flattened, armoured body and show that the limbs were paddle-like, with a curved upper arm bone resembling that of sauropterygians. Broad, closely packed gastralia form a bony basket on the underside. The skull is intermediate in proportions, being neither long- nor short-snouted. The nostrils are close to the eyes, the upper temporal openings are closed, and the cheek region is ventrally embayed. The teeth are unusual. Those at the front of the upper jaw have rounded bases that are wider than the pointed, curved crowns, while those lining the jaws have leaf-shaped crowns that narrow, widen, then narrow again.

These features are taken to an extreme in *Sinosaurosphargis*, the most turtle-like of this group. *Largocephalosaurus* is less specialized. It has a longer snout, narrower body and longer tail than *Sinosaurosphargis* and upper temporal openings are present in the skull.

How do saurosphargids relate to other marine reptiles? While turtle-like or placodont-like in their wide, flattened body and armour covering, saurosphargids lack features unique to those groups and do not appear closely related to them. They are, however, sauropterygian-like in the shape of their limbs and in how their clavicles articulate with the surrounding bones, so it seems likely that they shared an ancestor with sauropterygians. An affinity with thalattosaurs has also been suggested.

In view of this, saurosphargids likely evolved from a long-tailed, fish-eating ancestor that had a long, narrow body and long, narrow skull. Why they became broad-bodied, armoured animals with peculiar teeth and a shortened snout is an interesting question. One possibility is that they became bottom-feeders, adapted for a diet of reef-dwelling invertebrates. Bottom-feeding requires that animals become slower, and thus more vulnerable to predators. They also benefit by being heavy, since they are then able to remain at depth more easily.

A few other animals appear allied to saurosphargids. One of the most peculiar Mesozoic marine reptiles is *Helveticosaurus*, named in 1955 from a skeleton from Monte San Giorgio. Originally interpreted as close to placodonts, it was clearly something else. *Helveticosaurus* was reasonably large at around 4 m (13 ft) in length. It had a long body and short neck, and its long, flexible tail, powerfully muscled pectoral girdle and long, flipper-like forelimbs suggest a style of swimming that combined sculling with the tail and rowing or flapping with the limbs. The skull is short-snouted and deep, with massive eyes and lengthy fangs.

Helveticosaurus must have occupied a very particular lifestyle, the problem being that we have no idea what that lifestyle was. Perhaps it was an amphibious predator that lurked in shallow water and grabbed fish and small reptiles (maybe even from the land), or perhaps it was a manoeuvrable aquatic predator, good at twisting and turning in pursuit of fishes, cephalopods or small marine reptiles. The possibility exists that *Helveticosaurus* was doing something else entirely, such as tearing at marine algae, biting at the tips of limbs or tails of other reptiles, or consuming sea jellies or other soft prey. Some of those ideas could be tested through examination of the one known

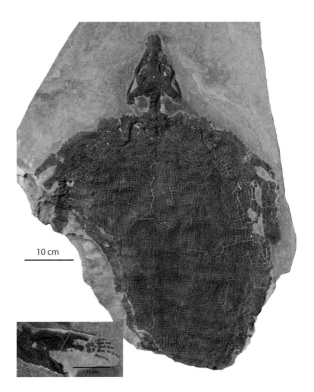

The saurosphargid *Sinosaurosphargis* of Yunnan and Guizhou Province, China. The upper surface of the broad, flattened body is covered by a flexible shield formed of hundreds of square and rhomboidal osteoderms. The forelimb is streamlined and osteoderms cover more than half of its length.

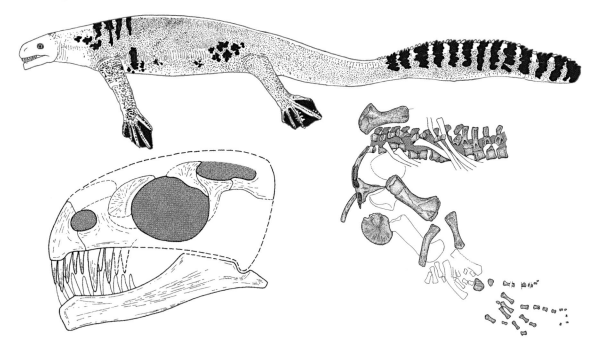

Helveticosaurus is an unusual animal about which little is known. Its skull – though imperfectly known – is short-snouted and with massive fangs. The fangs are so long that they were likely visible even when the jaws were closed. Long, probably webbed digits were used in swimming in addition to the long tail.

specimen, but others can't be checked until we have additional specimens that provide more information. Indeed, part of the problem is that we only have one individual. A pelvis from a *Helveticosaurus*-like animal is known from the Lower Triassic of Spitsbergen and probably represents a distinct genus.

Also worth mentioning here is *Eusaurosphargis*, named from Italy in 2003 but since reported from the Netherlands and Switzerland. *Eusaurosphargis* was short-tailed, deep-snouted, short-faced, and with short hands and feet. Triangular osteoderms on the sides of its broad body create serrated edges that perhaps provided protection from predators or from combat with its own kind. None of these features are typical for aquatic animals, and imply that *Eusaurosphargis* was terrestrial, or at least amphibious. Its size is uncertain since the only articulated specimen is a juvenile 20 cm (8 in) long, but adults were probably around 1.5 m (5 ft) long.

Eusaurosphargis does not appear to be a saurosphargid, but is outside the clade that includes saurosphargids and sauropterygians. It might be close to *Helveticosaurus*. Whatever its affinities, it is embedded within the superclade. If it really was terrestrial, one possibility is that it shows how one lineage within the superclade secondarily adapted to life on land. Alternatively, perhaps this lineage retained amphibious habits from ancestral members of the superclade that never became fully aquatic.

MARINE RHYNCHOCEPHALIANS

Among the most familiar of reptiles are the squamates, the vast group that includes lizards and snakes. Less familiar are the related rhynchocephalians (which means 'beak-heads'), today represented by a single species, the tuatara, *Sphenodon punctatus*. This is a chunky, lizard-shaped reptile 30 to 80 cm (12 to 31½ in) long, and unique to New Zealand. Tuataras prey on insects and spiders, though frogs, lizards and seabird chicks are sometimes also eaten. The name beak-head refers to the presence of paired, hooked, beak-like teeth at the tip of the upper jaw. These are obvious in the tuatara skull, albeit not in the live animal, and are a typical feature of fossil rhynchocephalians.

The tuatara's main claim to fame is that it is a living fossil: the lone survivor of a once widespread group, and a primitive animal, similar to ancient reptiles of the Triassic. An improved understanding of rhynchocephalian history and tuatara biology has shown that none of this is really correct. Those features of the tuatara thought to be 'primitive' are recently evolved novelties linked to its shearing bite and island-dwelling life on temperate New Zealand.

Rhynchocephalians were widespread and diverse from the Middle Triassic to the end of the Cretaceous, often occupying roles later filled by lizards. Many were insectivores but some were herbivores. Some Jurassic and Cretaceous species were specialists with teeth adapted for crushing mollusc shells, or had a covering of protective osteoderms or a venom-delivery apparatus in the lower jaw evolved. Acrodont teeth (meaning that the teeth are fused to the bones of the jaws) are typical in the group. Of interest to us here is that rhynchocephalians took to marine life on perhaps three occasions. None evolved giant size (the biggest were about 1.5 m or 5 ft long), and all adapted to the sea in different ways.

The least specialized of these are the sapheosaurs of Late Jurassic western Europe, fossils of which come from marine, Tethyan rocks. The best known of them – *Sapheosaurus* from Germany and France – reached 70 cm (27½ in) and its proportions are not especially different from those of terrestrial rhynchocephalians, though its tail is longer. Its jaws are toothless, raising questions of what it ate and how it ate it. Related animals, including *Kallimodon* and *Leptosaurus*, are from southern Germany and differ in possessing teeth. Those at the front of the jaws are rounded cones while those at the back are large and triangular.

The long, slender, aquatically adapted pleurosaurs of the European Jurassic – like *Pleurosaurus*, shown here – were extremely different in shape and lifestyle from the only living member of Rhynchocephalia, the Tuatara of New Zealand.

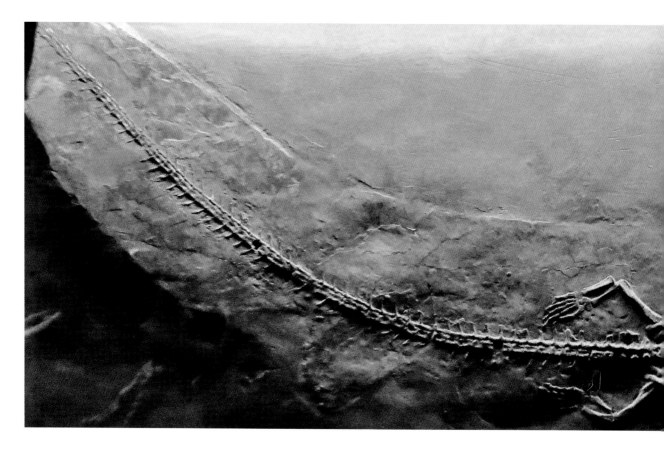

The most specialized marine rhynchocephalians are the pleurosaurs, a Jurassic group again best known from the Late Jurassic Tethyan sediments of Germany and France, though one early member (*Palaeopleurosaurus*) is from the Early Jurassic. Pleurosaurs are long-snouted, extremely long-bodied and long-tailed, and with limbs that are proportionally tiny in the later members of the group. They were clearly anguilliform swimmers. The acrodont teeth are triangular and mostly broad-based. The hooked teeth at the upper jaw tip form a large, curved 'beak' in *Palaeopleurosaurus*. Some *Pleurosaurus* specimens preserve segments of scaly skin. The scales are hexagonal with a tidy, geometrical shape and an enlarged row of hexagonal scales fringes the upper edge of the tail.

A possible early pleurosaur – *Vadasaurus* from the Late Jurassic of Solnhofen – has 'normal' proportions relative to the others, but its long tail, elongate nostrils and reduced ossification (see Chapter 3) suggest marine habits. Some studies find pleurosaurs to be close allies of sapheosaurs, implying that both groups share a marine ancestor.

Finally, we come to the third marine lineage among the rhynchocephalians: *Ankylosphenodon* from the Early Cretaceous of Mexico. *Ankylosphenodon* is large for a rhynchocephalian (around 1.5 m or 5 ft long), has a robust, dense-boned skeleton, comes from marine rocks, and has triangular, acrodont teeth that look suited for shearing but have wear suggestive of an herbivorous diet. Its limbs are well developed, though reduced ossification is present, and there are no obvious swimming adaptations. It is very tempting to suggest

This complete skeleton of *Pleurosaurus goldfussi* emphasizes the pointed snout, long body and extremely long tail of these animals. Pleurosaurs are so odd that they have sometimes been thought to represent an entirely new reptile group, termed Pleurosauria.

that *Ankylosphenodon* was a Cretaceous analogue of the modern Galápagos marine iguana: a capable swimmer, good at walking on land and clinging to submerged surfaces, and adapted for a diet of marine plants. Exactly this view of the animal was favoured by its describer, herpetologist Victor-Hugo Reynoso. Herbivory is rare in Mesozoic marine reptiles and rare in marine vertebrates in general, so this is significant.

CRETACEOUS SEA SNAKES

The best-known Mesozoic marine squamates are the mosasaurs, the predatory lizards covered in Chapter 8. Mosasaurs were not the only squamates that took to marine life during the Mesozoic: some snake groups did as well. The most interesting of these are the pachyophiids or simoliophiids (both names are currently in use), all of which are from the middle of the Cretaceous, between 100 and 93 million years ago. They are unique to Europe, north Africa and the Middle East. Three points make pachyophiids of special interest: they were marine, they possessed hindlimbs, and their anatomy means that some experts interpret them as especially closely related to the snake common ancestor. If that last point is true, they are of crucial importance in understanding snake origins.

The first pachyophiid to be described was *Simoliophis* from France, Portugal and north Africa, named from vertebrae in 1880. A better-known animal – *Pachyophis* – was described from Bosnia-Herzegovina in 1923,

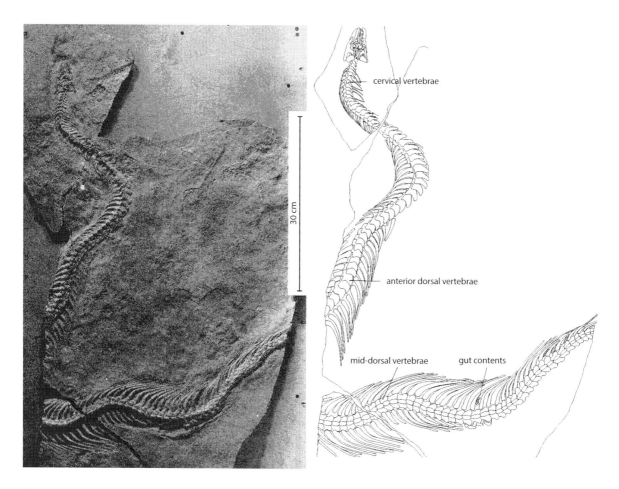

The marine Cretaceous snake *Pachyrhachis* is known from two near-complete skeletons. Various of its features are suggestive of a sluggish, slow-moving lifestyle. Around 150 vertebrae formed the neck and body (the tail of *Pachyrhachis* is unknown), and all are short, broad and chunky.

but not fully appreciated as a Cretaceous snake until the discovery of *Pachyrhachis* from the Jerusalem area of the Middle East in 1979. Later came *Haasiophis* from Jerusalem and *Eupodophis* from Lebanon.

The pachyophiid skull is proportionally small but is similar to that of living pythons. Recurved teeth line both the jaws and certain bones on the palate, and the jaws could be opened wide due to movement at the jaw joint, mid-way along the lower jaw and at the chin, too. The skeleton has thickened, dense bones. The most surprising feature is tiny hindlimbs, complete with knee and ankle joints and tiny toe bones. These legs are remarkably small and located extremely far back. Pachyophiids are mostly 1 to 2 m (3¼ to 6½ ft) long, yet their hindlimbs are only about 2 cm (¾ in) long.

The marine lifestyle of these snakes is demonstrated by the sediments in which their fossils are found, and the fact they had dense bones. But other clues link to their seagoing habits. Their proportions are different to land snakes. Snakes mostly consist of a long body, the neck and tail being short, but in pachyophiids, the tail is not just short, but laterally compressed, and so is the rear of the body. Overall, pachyophiids look like slow-moving animals of shallow water. They might have hunted prey in burrows and crevices, or perhaps they waited on the seafloor and grabbed fish with a fast strike. One *Pachyrhachis* specimen has the remains of a fish within its gut.

The presence of hindlimbs in pachyophiids suggests that they are especially early members of the snake lineage, close in evolutionary terms to the lizards that were snake ancestors. If this is true, and given that these animals were marine, here is evidence that snakes originated in the sea. Some experts argue that snakes and mosasaurs are closely related, together forming a clade that evolved its unusual features after taking to marine life.

Other experts disagree with this proposal. Snakes possess features suggestive of a terrestrial origin that involved burrowing and digging. Furthermore, limblessness in squamates is overwhelmingly linked with a burrowing, terrestrial life, not a swimming one. The presence of hindlimbs in pachyophiids makes them look 'primitive' relative to other snake groups, but their other features – recall their python-like skulls – are typical of the evolutionarily young snakes called the macrostomatans. Some snake experts therefore argue that pachyophiids are not close to snake origins at all, but are instead members of the macrostomatan group.

If this is true, perhaps the presence of hindlimbs was widespread in Cretaceous snakes, and perhaps pachyophiids were among several groups that retained them. Furthermore, a macrostomatan position for pachyophiids would mean that their marine habits were specialized and novel, and not relevant to snake origins. What pachyophiids mean for snake origins, where snakes belong within the squamate family tree and how and where snakes originated remain among the most debated issues in fossil reptile research.

The limbs of pachyophiids like *Pachyrhachis* here were so small and slender that they were probably hard to spot in the live animal. Pachyophiids appear to have inhabited reefs and other cluttered environments, so were probably camouflaged with spots or stripes, as inferred here.

5 | SHARK-SHAPED REPTILES:
THE ICHTHYOSAURS AND THEIR KIN

THE ICHTHYOSAURS, SOMETIMES CALLED 'fish lizards', are among the most recognizable of the Mesozoic marine reptiles. Articulated skeletons show that most were streamlined and shark-shaped, and usually between 2 and 6 m (6½ and 19¾ ft) in length. A few giants reached 20 m (65½ ft) and more. Typical features include long, slender jaws, conical teeth, enormous eyes, two pairs of fin-like limbs, a dorsal fin and a crescent-shaped tail fin. This description applies to the members of Parvipelvia, the group that originated at the end of the Triassic and persisted until the middle of the Cretaceous.

Several non-parvipelvian lineages evolved during the Triassic. These were not as anatomically homogenous as parvipelvians and evolved in several different directions. Some had blunt-tipped or rounded teeth, while others were large, even enormous, predators. All, however, were built for life in water and were devoid of features that might allow movement on land.

The absence of fossils that look like proto-ichthyosaurs – that fill the gap between these aquatic species and their terrestrial ancestors – has left experts in the dark on ichthyosaur origins, and on what the earliest stages of their evolution were like. Recent discoveries have changed that. Several proto-ichthyosaurs are now known, including the hupehsuchians and nasorostrans of the Early Triassic. All are from China, suggesting that ichthyosaurs originated in the Eastern Tethys Ocean. In addition, new

We mostly think of ichthyosaurs as dolphin-sized animals. On several occasions, however, they gave rise to gigantic predators, some of which exceeded 15 m (50 ft). This giant is *Temnodontosaurus* from the Early Jurassic.

specimens of Triassic ichthyosaurs such as *Utatsusaurus* from Japan have shown that ichthyosaurs almost certainly descended from ancestors with a typical diapsid skull architecture.

The fact that ichthyosaurs of several groups are known from the oldest part of the Early Triassic shows that their evolution was well underway by this time, just five million years after the extinction event that ended the Permian, 252 million years ago. They evolved extremely quickly, reaching giant size in that relatively short time.

ICHTHYOSAUR DIVERSITY AND HISTORY

Ichthyosaur history can be roughly broken down into three stages. The first involves the earliest, most anatomically archaic kinds, the grippidians and other early forms of the Early Triassic. These are mostly small, 1 to 3 m (3¼ to 10 ft) long, and had elongated limb bones with proportions and shapes that hint at terrestrial ancestry. The second stage includes the larger, more predatory cymbospondylids and shastasaurs, species of which were present from early in the Triassic until its end. These included the largest ichthyosaurs. Elongated snouts and jaws are typical of these animals, as are enlarged, narrow limbs. Some shastasaurs evolved into those ichthyosaurs belonging to the third stage, the parvipelvians, the shark-shaped ichthyosaurs associated mostly with the Jurassic. Parvipelvians include the most streamlined and pelagic of ichthyosaurs, the thunnosaurs and their kin, and their key features include extra digits, reduced hindlimbs, a reduced pelvis, and a short tail where the end of the spine is sharply downturned to

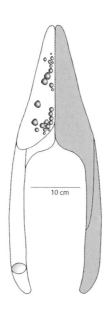
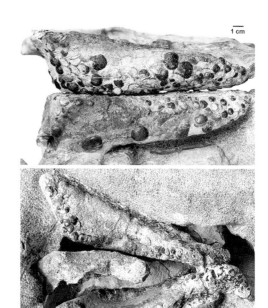

These jaw bones, peppered with button-shaped teeth, belong to *Omphalosaurus*. The image on the left shows a reconstructed lower jaw showing the upper surface on the left and lower surface on the right. The two images on the right show the underside of the bones that form the tip of the upper jaw (top), and the upper surfaces of the front parts of the lower jaw (bottom).

support the lower lobe of a crescent-shaped tail fin. Their reduced pelvis explains their group name, Parvipelvia meaning 'small pelvis'.

Certain Triassic ichthyosaurs are especially odd and cannot be placed within the major groups. In some cases, their ichthyosaurian status has even been doubted. Among these are the omphalosaurs of Early and Middle Triassic central Europe, Spitsbergen and the USA. A pavement of button-like teeth is present across their upper and lower jaws and one species had a spatulate lower jaw.

According to some experts, grippidians and their kin are different enough from more typical ichthyosaurs to be excluded from the group Ichthyosauria. These experts prefer to use the name Ichthyopterygia for this grippidian + ichthyosaur clade, the result being that grippidians are 'non-ichthyosaurian ichthyopterygians' according to this view! Other experts disagree and include grippidians within Ichthyosauria. Whichever of these views is preferred, the majority of experts refer to all ichthyopterygians as ichthyosaurs, and that convention is followed here.

Numerous trends are seen in ichthyosaur evolutionary history. The most obvious involves the transition from anguilliform swimming (using undulations of the body and tail) to carangiform (tail undulations only) and ultimately thunniform (the fastest, using just the tail fin), as described in Chapter 3. Ichthyosaurs retained both limb pairs throughout their evolution, even though the hindlimbs of parvipelvians were much reduced. Other trends include a reduction of the skull behind the eyes and a change to the scapula from a fan-shaped plate to a slender blade with a broad base.

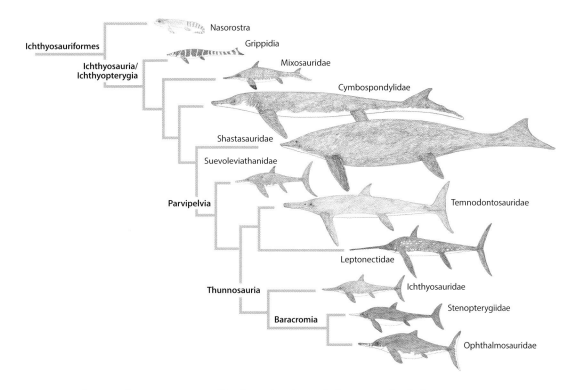

Until recently, it was generally thought that ichthyosaurs had their heyday in the Triassic and Early Jurassic and persisted at low diversity from the Middle Jurassic until their extinction in the middle Cretaceous. A huge number of studies and discoveries have overturned this view. They show that parvipelvians thrived at diversity in the Late Jurassic and Early Cretaceous, the group finally waning around 100 million years ago at the end of the Early Cretaceous. Virtually all of these late-surviving forms belong to the same group, Ophthalmosauridae, so it is still true that this final flowering of the group did not fill as many roles in ecosystems as did the ichthyosaurs of the Triassic. The final, middle Cretaceous extinction of the group was related to an overturn that happened in marine ecosystems at this time (see Chapter 2).

The ichthyosaur family tree shows how ichthyosaurs evolved several different body shapes early in their evolution but were more conservative in shape later on. The majority of familiar ichthyosaur belong to Parvipelvia.

This graph, published by ichthyosaur expert Valentin Fischer in 2011, shows how new discoveries (in purple) increased the number of Early Cretaceous ichthyosaurs relative to earlier thinking (the Cretaceous starts with 'Ber', short for Berriasian). A flurry of even newer discoveries mean that this graph is now quite out of date!

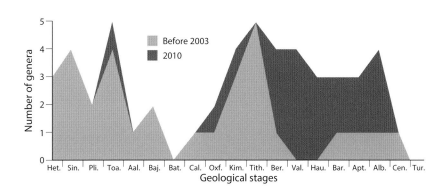

NASOROSTRANS

It appears that hupehsuchians, discussed in Chapter 4, are the closest fossil animals we have to ichthyosaur ancestors. Aside from these, the oldest and most archaic members of the ichthyosaur lineage are two Early Triassic reptiles from China, both discovered during the twenty-first century. These are the small *Cartorhynchus*, which was 40 cm (15¾ in) long, and the larger *Sclerocormus*, which was 1.6 m (5¼ ft) long. Both share features of the skull and ribcage and are united in the clade Nasorostra, which means 'nose snout'.

The suggestion has been made that *Cartorhynchus* was not restricted to aquatic life but was also capable of movement on land. Its limbs appear to have been flexible, perhaps suggesting they were used to support its weight, and it is small enough to have been capable of movement in shallow water or at the water's edge. Its skull suggests a lifestyle different from that of ichthyosaurs proper. The short snout, deep face, enlarged, heavily built hyoid bones in the throat and rounded teeth indicate that it was a suction feeder, able to vacuum animals such as crustaceans and shellfish into its mouth. *Sclerocormus* also has a short snout but its head is even smaller, proportionally, and its body is stiffer, with broader, flatter ribs.

Based on the anatomy of thalattosaurs, hupehsuchians and early ichthyosaurs, it seems sensible to assume that ichthyosaurs evolved from long-jawed animals equipped with pointed teeth. The short-snouted nasorostrans

If – as experts suggest – *Cartorhynchus* was amphibious, it might have used terrestrial environments in courtship and mating. This reconstruction depicts an individual moving across a gravel bar at the edge of an estuary.

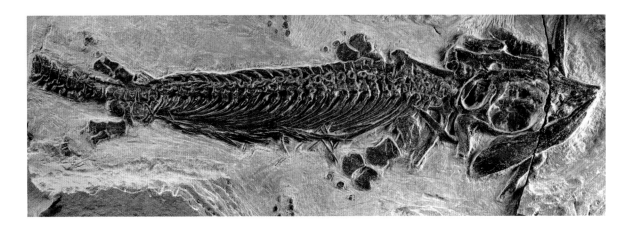

are not consistent with that. Perhaps they show that ichthyosaurs originated as nasorostran-like suction-feeders, or perhaps nasorostrans are a side-branch that evolved from less unusual, longer-jawed ancestors.

GRIPPIDIANS, CHAOHUSAURS AND KIN

In 1929, an unusual, surprisingly old ichthyosaur was discovered in the Lower Triassic of Spitsbergen. Named *Grippia* by Swedish palaeontologist Carl Wiman, in honour of German geologist Karl Gripp, it was peculiar enough that Wiman didn't realise its ichthyosaur affinities. In contrast to other ichthyosaurs known at the time, *Grippia* was small, around 1 m (3¼ ft) long, had small teeth with rounded, blunt tips, and possessed limbs where the bones were similar to those of terrestrial reptiles. The snout was missing, but what was preserved suggested that the skull was triangular when seen from the side. This was not a speedy, pelagic animal, but a slower-moving denizen of shallow water. Later discoveries in China, Japan, Thailand and Canada showed that *Grippia* was not an oddity, but typical of Early Triassic ichthyosaurs. Several share features and are united within a group termed Grippidia.

At 3 m (10 ft) in length, *Utatsusaurus* is the biggest of these early ichthyosaurs. In its skull, the arrangement of bones around the upper temporal opening is the same as it is in diapsid reptiles, providing support for the inclusion of ichthyosaurs within this group. Elsewhere in the skeleton, two bony ribs connect the pelvis to the spine (a feature associated with weight-bearing, and normally absent in aquatic animals), and the hindlimbs are slightly larger than the forelimbs. These animals might be archaic relative to parvipelvians, but they were still fully aquatic, with flippered limbs, a streamlined skull, and a tail where tall

The skeleton of *Cartorhynchus* reveals the proportionally large skull characteristic of this animal. The skull looks toothless but CT-scanning has revealed the presence of numerous rounded, molar-like teeth in both the upper and lower jaws.

As is typical for marine reptiles, *Grippia* has large eye sockets and 'retracted' nostrils located well away from the snout's tip. When complete, the snout and jaws of this skull would have extended much further forward than preserved here. The sclerotic ring is missing from this fossil.

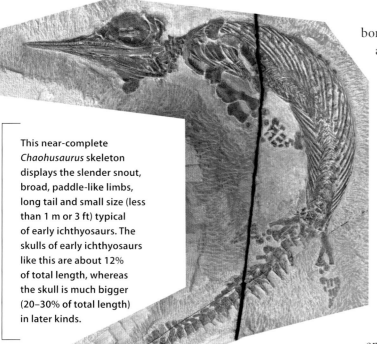

This near-complete *Chaohusaurus* skeleton displays the slender snout, broad, paddle-like limbs, long tail and small size (less than 1 m or 3 ft) typical of early ichthyosaurs. The skulls of early ichthyosaurs like this are about 12% of total length, whereas the skull is much bigger (20–30% of total length) in later kinds.

bony spines in the end half supported a vertical tail fin. The teeth at the front of their jaws are slim and pointed, while those further back are blunter and rounder. These teeth are not suited for the breaking of especially hard food items, but more likely used in the crushing of all sorts of prey, including thin-shelled molluscs, crustaceans and fishes.

Chaohusaurus from China, a close relative of grippidians, is known from numerous specimens that represent more than one species. These differ in flipper shape. This suggests that they were occupying different environments. One *Chaohusaurus* specimen is preserved with a dorsal fin, though this is difficult to make out and could just be a mark on the rock surrounding the fossil. Another specimen contains three unborn young. This is the oldest evidence for viviparity in ichthyosaurs. One of the young is projecting head-first, the opposite direction from normal in later ichthyosaurs. Presumably, head-first birth was normal early in ichthyosaur evolution, and only later evolved into tail-first birth.

The distribution of grippidians, chaohusaurs and other early forms requires their presence in both eastern and western Panthalassa. A cross-Panthalassa swim seems unlikely for such small, mostly coastal swimmers, so experts have suggested that they used coastal routes to disperse around the world. Maybe they occurred right around Pangaea's north. The possibility that they travelled around the south of Pangaea has also been suggested. A final possibility is that they used a channel that traversed northern Pangaea and formed a connection between the Tethys Ocean and eastern Panthalassa. Grippidians and their kin had essentially died out by around 240 million years ago. However, at some point they gave rise to the group that includes all later ichthyosaurs.

MIXOSAURS

By the latter half of the nineteenth century, it was agreed that ichthyosaurs older than, and structurally more primitive than, the parvipelvians of the Jurassic had yet to be discovered. The first of such animals known to science was announced from the Middle Triassic rocks of Germany in 1852, and is called *Mixosaurus*. Discoveries made since 1852 show that *Mixosaurus* and its relatives, formally termed Mixosauria, are only distantly related to parvipelvians. This wasn't obvious when they were found and they were initially misidentified as Triassic species of *Ichthyosaurus*. During the 1880s, German zoologist and palaeontologist Georg Baur realised – thanks mostly to flipper structure – that mixosaurs are substantially more archaic than parvipelvians. Today we recognize them as the oldest members of the group that evolved from the grippidian-like forms of the Early Triassic.

Mixosaurs are mostly small, 1 to 2 m (3¼ to 6¾ ft) long and lack the crescent-shaped tail fin present in parvipelvians. They possess a large, fan-shaped scapula, as is typical of Triassic ichthyosaurs. In the limbs, the bones of the lower arm and lower leg are approximately rectangular, separated by prominent gaps, and distinct from the bones of the wrists and hands, and ankles and feet. In these respects, mixosaurs have a more 'ancestral' look relative to post-Triassic ichthyosaurs. They do, however, possess some advanced features, including a round eye socket and shortened limb bones. These details place them closer to shastasaurs and parvipelvians than other Triassic ichthyosaurs.

A few features are peculiar to mixosaurs. Their neural spines (the bony rods projecting upwards from the tops of the vertebrae) are tall, and the vertebrae in the middle of the tail are large relative to the others. These features presumably relate to swimming style and might show that mixosaurs had an unusually muscular spine and mid-tail region. Perhaps they were capable of impressive bursts of speed. Given that they were generally small compared with other predatory marine reptiles of the time, such as shastasaurs and giant nothosaurs, we might speculate that this was used as an escape response.

One *Chaohusaurus* specimen is preserved with embryos inside its body. Several Jurassic ichthyosaurs are preserved with their embryos arranged such that the baby would have emerged tail-first at the moment of birth. *Chaohusaurus* was not like that – it gave birth in the 'head-first' way, as depicted in this reconstruction.

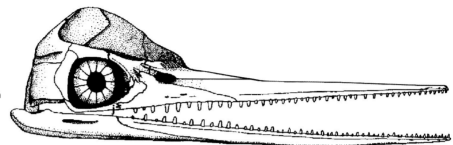

The reconstructed skull of *Contectopalatus*, a controversial giant mixosaur from Germany and China. Its conical teeth look suited for the capture of cephalopods and fish. At least some of the tooth crowns have vertical striations on their sides and blunted tips.

The mixosaurian snout is long, slender and shallow relative to the cranial region, the eye sockets are huge, the bony palate is closed and complete (more usually in ichthyosaurs, a V-shaped gap divides the palate) and the teeth are mostly conical. Their teeth belong to the 'smash' or 'crunch' guild (see p. 40) and indicate they were predators of squid-like molluscs or unarmoured fishes, an idea backed up by stomach contents. The species of *Phalarodon* have mound-shaped teeth at the backs of their jaws in addition to conical teeth at the front. The rounded teeth look suited for crushing, so we presume that these species exploited such prey as crustaceans and thin-shelled molluscs. They probably weren't cracking open especially tough-shelled prey, since their jaws were slender.

More unusual is the sagittal crest, the tall, narrow bony crest that projects from the forehead, sometimes extending forward to the base of the snout. Large, concave areas are present on either side. One suggestion is that large muscles were attached to the crest, so perhaps it helped provide a powerful bite. The crest increased in height with age, so the possibility also exists that it functioned as a sign of maturity.

The streamlined shape of the mixosaur body suggests the presence of a dorsal fin. An Italian *Mixosaurus* specimen, published in 2020, confirms dorsal fin presence, but it was more slender and positioned further forward than suspected. It would have been supported internally by stiffening fibres. Both its stiffness and position over the deepest part of the ribcage indicate it had a role as a stabilizer.

After their discovery in Germany, mixosaurs were reported from Italy, Spitsbergen, Switzerland, Poland, Russia, Canada and the USA, and more recently from Turkey, Indonesia and China. The majority known so far are agreed to represent just two genera: *Mixosaurus* and *Phalarodon*. A few finds, however, indicate that mixosaurs were more diverse than usually implied. *Wimanius* from Switzerland is unusual in lacking a sagittal crest and in having a row of small teeth on its palate. Mixosaur remains from Germany and China hint at the presence of especially big specimens where the skull was as long as 80 cm (31½ in), suggesting a total length of 5 m (16½ ft). These large mixosaurs had a long, extremely narrow snout and especially tall sagittal crest. They are recognized by some experts as the taxon *Contectopalatus*, but by others as large specimens of *Mixosaurus* or *Phalarodon*.

CYMBOSPONDYLIDS

In 1868, a new ichthyosaur group was recognized for Middle Triassic fossils from Nevada. These are the cymbospondylids. This awkward name is based on that of the best known member of the group: *Cymbospondylus*. The concave front faces of its vertebrae explain its name, which means 'cup-like vertebra'. Cymbospondylids include small species around 2 m (6½ ft) long as well as giants of 17 m (55¾ ft). Long, robust jaws and large muscle attachment sites at the back and top of the skull are typical, but the teeth are small and look suited for fish and cephalopod prey. The small *Xinminosaurus* is different, its bulbous teeth and toothless snout tip suggesting a diet of shelled invertebrates.

Several species of *Cymbospondylus* are known from the USA. One Monte San Giorgio specimen is regarded by some experts as a European *Cymbospondylus* but by others as representing the distinct genus *Phantomosaurus*, its name being a reference to the protagonist of the 1911 novel *Phantom of the Opera*. The reason for this name is that the fossil's face was damaged by careless treatment with acid!

Cymbospondylid proportions are unusual. The skull and limbs are small relative to overall length, the body is long and shallow, and the tail preserves evidence for a low vertical fin. It looks almost as if an archaic ichthyosaur such as *Chaohusaurus* has been 'maximised' to the greatest size possible, and it might be correct to imagine cymbospondylids as the earliest evolutionary effort by ichthyosaurs to reach giant size.

The scapula in some cymbospondylids is odd, being vaguely shaped like a boot, with the 'heel' forming a huge contribution to the shoulder joint. Ordinarily in Triassic ichthyosaurs, the scapula is semi-circular, with a small section on its trailing margin contributing to the shoulder joint. The coracoid is odd, too, with a concave trailing margin and an opening for nerves and blood vessels. These features are related to forelimb musculature and function, but what they might mean is unstudied. Another odd cymbospondylid feature is that the part of the skull that articulated with the neck is concave, not convex as is typical. We have no good idea why this is, but it might be a throwback to the condition present in ancestral reptiles. The cymbospondylid eye socket is oval rather than round and the aperture in the middle of the sclerotic ring is small. This means that the pupil would have been smaller than it was in parvipelvians. However, it was still of a size consistent with excellent vision comparable to that of birds.

The cymbospondylid body shape could mean that they were pelagic animals of surface waters. This is consistent with their distribution, since fossils have been found in the east of Panthalassa, the Boreal Ocean and in the Western and Eastern Tethys Ocean. This suggests a global distribution, or at least one around the north, east and west of Pangaea. Some cymbospondylid fossils from the USA and Spitsbergen are Early Triassic in age, which is

consistent with an explosive diversification of ichthyosaurs close to the start to the Triassic, and perhaps with Permian origins.

One cymbospondylid specimen from Nevada has the skeletons of three young within its body. This is the oldest evidence of ichthyosaur pregnancy after the Early Triassic *Chaohusaurus* specimen mentioned earlier. These unborn young are (relative to the size of the mother) similar in size to those of *Chaohusaurus* but 9% larger than those of parvipelvians such as *Stenopterygius*. This suggests that ichthyosaur young became proportionally smaller over the course of evolution.

In some respects, cymbospondylids are similar to another group of mostly large ichthyosaurs, the shastasaurs, and both have sometimes been considered close relatives. However, shastasaurs possess features that link them with parvipelvians. Cymbospondylids do not.

SHASTASAURS

For more than 30 million years, Triassic seas were inhabited by a group of mostly long-jawed, long-bodied, mostly large and sometimes gigantic ichthyosaurs known as the shastasaurs. The biggest species were among the largest animals of all time. A Canadian shastasaur reached at least 21 m (69 ft) while remains from England and France (in particular jaw bones from Aust in England) suggest lengths of 25 to 30 m (82 to 100 ft) and possibly more. This globally distributed Middle and Late Triassic group evolved a diversity of lifestyles. Shastasaurs include the ancestors of the one ichthyosaur group that persisted beyond the Triassic (the parvipelvians) and are thus especially important in ichthyosaur history.

Our knowledge of shastasaurs began in 1895, when American palaeontologist and conservationist John C Merriam described *Shastasaurus* from the Upper Triassic of Shasta County, California. In the forelimb, the upper and lower arm bones are block-like and only three fingers are present. A partial skull was described in 1902. Its eye sockets are oval, the temporal openings long and large, but the snout is broken off. Merriam assumed that the complete skull was long-snouted and long-jawed, like that of *Ichthyosaurus*. He referred to conical teeth with striated crowns, discovered in the same rocks, and reasoned that they belonged to *Shastasaurus*.

During the late twentieth century, additional shastasaurs were reported from the Upper Triassic of China, the USA, Italy, Austria, Germany and Canada. The twenty-first century has seen the discovery of Chinese shastasaurs as well as the realization that finds from France – first published in the early 1900s – also represent shastasaurs, and demonstrate the group's persistence to the end of the Triassic.

The most impactful of these discoveries was *Shonisaurus popularis*, initially reported from Berlin, Nevada (USA, not Germany) in 1976. This was a huge ichthyosaur with a long, heavily built snout and long, tridactyl

This reconstruction depicts the gigantic cymbospondylid Cymbospondylus petrinus *from Nevada, USA, named in 1868. Its skull is 1.1 m (3½ ft) long, yet some cymbospondylid species were even larger. This image hints at the idea that ammonites were among its favoured prey.*

Several shastasaurs (this is *Guanlingsaurus* from China) were toothless and short-snouted relative to other ichthyosaurs. They were mostly large – *Guanlingsaurus* was 10 m (33 ft) long. A near-complete *Guanlingsaurus* juvenile, described in 2013, shows that the body was extremely long. It contained around 80 vertebrae, about twice as many as are typical for ichthyosaurs.

flippers. Numerous other specimens have since been discovered, all nearby. Several have been left in place at the discovery site and today form the central display items in Berlin Ichthyosaur State Park, perhaps the world's only tourist feature devoted to ichthyosaurs. It remains unknown why these animals are preserved together. One suggestion is that they became stranded on a beach together, much as whales do today. Another is that the specimens were only preserved together by chance, each drifting to the seafloor at a different time.

Merriam's original *Shastasaurus* indicates a length of 7 m (23 ft). *Shonisaurus popularis*, in contrast, was a giant, some specimens reaching 15 m (49 ft). The giant shastasaur from British Columbia, mentioned above, is even larger. Its skull, though incomplete, is 3 m (10 ft) long, and the one specimen excavated so far has an estimated length of 21 m (69 ft). However, the describers of this animal noted that bones from even larger individuals have been found at the same location. When first described, *Shonisaurus* was thought to be unusually shaped, with a deep belly and short tail. Recent studies show that it was instead long-bodied and long-tailed, much like other shastasaurs.

Himalayasaurus is another giant from the Late Triassic of China. It also exceeded 15 m (49 ft) and has large, dagger-shaped teeth with prominent keels. The crown of each tooth is about 6 cm (2½ in) long, similar to crown size in the great white shark. *Thalattoarchon* from the Middle Triassic of Nevada is also large at 9 m (29½ ft and also has big, keeled, dagger-shaped teeth. These finds demonstrate that some shastasaurs were formidable predators.

In 2011, German palaeontologist Martin Sander and colleagues presented new data on the Chinese shastasaur *Guanlingsaurus*. They argued that it had a short snout, toothless jaws, and that its skull was proportionally small, less than 10% of the animal's length. Sander and colleagues described *Guanlingsaurus* as remarkably long in the body and tail, and with unusually short limbs. Here we come back to Merriam's original, broken *Shastasaurus* skull. Sander and colleagues argued that *Shastasaurus* wasn't lacking a long snout and jaws. Instead, they argued, it never had them. It was another short-snouted shastasaur, as was the giant from British Columbia. They grouped these animals together as species of *Shastasaurus*, changing the composition of this genus relative to traditional understanding. Sander and colleagues argued that these were long-bodied shastasaurs that used their short, toothless jaws in suction-feeding. Suction-feeding is common in aquatic animals and has evolved repeatedly among cetaceans, being well known in short-snouted animals such as beluga whales. Could it be, then, that *Shastasaurus* was specialized for suction-feeding, and therefore similar through convergent evolution to suction-feeding cetaceans?

As intriguing as this idea is, it has not stood the test of time. *Guanlingsaurus* lacks the features required for suction-feeding. It jaws are too long and narrow to generate suction, and the hyoids are not large enough to allow the expansion of the throat needed for suction-feeding either. By inference, the same was true for other short-snouted shastasaurs. Their real lifestyle remains enigmatic, but they weren't suction-feeding.

What sort of lifestyle were other, more conventional shastasaurs leading? Direct evidence comes from a specimen of the Chinese *Guizhouichthyosaurus* with a thalattosaur in its stomach. The thalattosaur was around 4 m (13 ft) long, and the shastasaur was 4.8 m (15¾ ft) long, though note that the thalattosaur was more lightly built than the shastasaur. Because most of the thalattosaur's body is present, with only the head and tail missing, it must have been swallowed in a single event, not in pieces. It may not be irrelevant that the ichthyosaur died before digesting this meal. Perhaps it choked to death or died from overeating. The fact, however, that *Guizhouichthyosaurus* could consume prey of this size is surprising since it is not an especially large ichthyosaur, nor one with teeth that look suited for the job. Its teeth put it in Massare's 'smash' guild, meaning it 'should' be a cephalopod predator. This fossil proves that the consumption of huge prey was happening in Triassic marine ecosystems, just as it was in the Jurassic and Cretaceous.

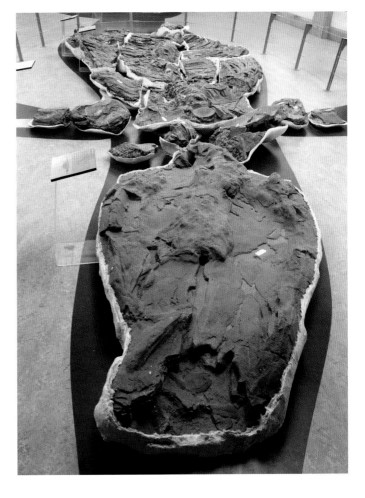

Between the mid-1990s and 2000, numerous giant shastasaurs were discovered in the Upper Triassic sediments of northeast British Columbia. Among the best of them is this enormous partial skeleton – 21 m (69 ft) long – collected on the banks of the Sikanni Chief River. Debate surrounds the identity of the Sikanni shastasaur. Originally labelled *Shonisaurus sikanniensis*, it was later argued to be a species of *Shastasaurus*.

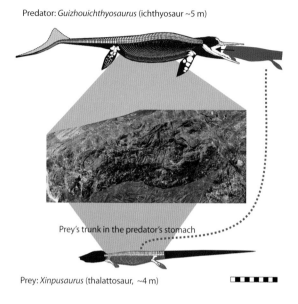

Predator: *Guizhouichthyosaurus* (ichthyosaur ~5 m)

Prey's trunk in the predator's stomach

Prey: *Xinpusaurus* (thalattosaur, ~4 m)

A *Guizhouichthyosaurus* specimen from China, discovered in 2010 but not scientifically described until 2020, contains the remains of the thalattosaur *Xinpusaurus* in its stomach. The thalattosaur lacks any evidence of being digested by stomach acids, so the ichthyosaur must have died soon after swallowing it.

Other shastasaurs were very different. Excessively slender jaws and tiny conical teeth are present in the large *Besanosaurus* from Italy, which grew up to 8 m (26 ft) long. This was mostly a predator of small, soft-bodied animals such as small fishes and squid. One *Besanosaurus* specimen contains small ichthyosaurs identified as embryos. However, they might be stomach contents. If so, they show that this shastasaur was capable of capturing and consuming prey again larger than those predicted from jaw and tooth anatomy.

How shastasaurs relate to other ichthyosaurs, and where they fit on the family tree, is controversial. In contrast with other Triassic ichthyosaurs, shastasaurs have a pared down shoulder blade, simplified radius and ulna, rounded rather than square or rectangular phalanges, and a shortened basket of gastralia which does not span the full distance between the pectoral and pelvic girdles. These features show that shastasaurs are closer to parvipelvians than the ichthyosaurs we've seen so far. While shastasaurs were traditionally grouped together, it has proved difficult to find details that unite them. What seems likely is that the term shastasaur applies to several lineages positioned between mixosaurs and parvipelvians on the family tree. Various names have been applied to these lineages, including Besanosauridae, Shonisauridae and Guanlingsauridae. A formal Shastasauridae is still recognized by some, the proviso being that it might contain *Shastasaurus* alone.

Shastasaurs are variable in anatomy and proportions, but those closest to parvipelvians are generally small, have an especially short region at the rear of the skull, a reduced maxillary bone on the side of the face, a reduced shoulder blade, narrow flippers and narrow neural spines. *Callawayia* from the Late Triassic of British Columbia and China possesses all of these features. At 2 m (3¼ ft) in length, it is the smallest shastasaur, and the most parvipelvian-like.

THE LATE TRIASSIC RISE OF PARVIPELVIANS

Around 230 million years ago, during the early part of the Late Triassic, a *Callawayia*-like shastasaur gave rise to a lineage of smaller, more streamlined, shorter-bodied ichthyosaurs. These include *Californosaurus* from (rather obviously) California, *Auroroborealia* from the New Siberian Islands in the Russian Arctic and the toretocnemiids from China and California.

Features that link these animals with parvipelvians include an even more reduced shoulder blade, a shorter tail and a more prominent tail bend. They might have been the first ichthyosaurs to possess the combination of a long-jawed, big-eyed skull, teardrop-shaped body, dorsal fin and crescent-shaped tail fin. We lack good information on their biology and behaviour, but it's conceivable they were faster swimmers, more efficient long-distance cruisers, or deeper divers than shastasaurs and other Triassic groups.

Perhaps one or more of these factors gave them an advantage during the extinction events at the end of the Triassic. Shastasaurs were killed off at this time, but these smaller kinds were not, and here is where one of the most important events in ichthyosaur history occurred: these small, Late Triassic forms were the ancestors of parvipelvians, the one ichthyosaur group that persisted into the Jurassic. Early parvipelvians of the Late Triassic – examples include *Macgowania* and *Hudsonelpidia*, both of British Columbia – were also small, their key innovation being a small pelvic girdle. Parvipelvians also possessed a steeper-angled tail bend than other ichthyosaurs, taller vertebrae where the bodies of the vertebrae are disc-shaped, and more tightly packed bones in the wrist.

SUEVOLEVIATHANIDS

As the Triassic ended and the Jurassic began, a turnover in the ichthyosaur lineages of the time is obvious, since all Jurassic ichthyosaurs belong to Parvipelvia. A surprising detail of early parvipelvian history is that the group did not merely originate prior to the end of the Triassic, it also diversified, since all four of its lineages (suevoleviathanids, temnodontosaurs, leptonectids, and thunnosaurs) persisted across the Triassic-Jurassic boundary. Of these four parvipelvian groups, only thunnosaurs had a future beyond the Early Jurassic.

Suevoleviathanids are a rare group, named for *Suevoleviathan* from the Posidonia Shale. This animal has been known since the 1840s but was variously misidentified as a species of *Ichthyosaurus*, *Temnodontosaurus*, *Leptonectes* or *Stenopterygius*. It was not until 1998 that German ichthyosaur expert Michael Maisch named it *Suevoleviathan*, meaning 'leviathan of Swabia' (Swabia is the old name for the region of southwest Germany that contains Holzmaden). Additional specimens have since been reported from southern France, but it remains rare: of around 3,000 Holzmaden ichthyosaur fossils, less than 10 are *Suevoleviathan*.

Suevoleviathan was 4 m (13 ft) long (and so not much of a leviathan) and possessed a low, streamlined skull where the jaws are neither especially robust nor slender. But the teeth are robust. Grooves on the bones of the snout might be linked to sensory abilities, but also perhaps to the presence of blood vessels or nerves. The most unusual feature of *Suevoleviathan* is its forelimbs. These are large and with digits that splay out at their tips. It also appears that the lower lobe of its tail was long and flexible. In one specimen the vertebrae at the tail's tip are preserved in a loop.

This *Suevoleviathan* specimen – kept at Stuttgart, Germany – displays the broad forelimbs typical of this ichthyosaur. The vertebrae at the tail's tip are preserved in a loop. This shows that the lower lobe of the tail was flexible.

The jaws, teeth and size of *Suevoleviathan* suggest it was an ecological generalist – its teeth put it in Massare's 'smash' guild. But its limbs and tail suggest it was adapted for foraging near the seafloor or around reefs or canyons, its large forelimbs and flexible tail enabling it to make constant small adjustments to its position. Those possibly sensory grooves on the snout are also consistent with this sort of lifestyle. An alternative possibility is that *Suevoleviathan* was an ambush hunter able to generate massive acceleration.

TEMNODONTOSAURS

Far better known than the suevoleviathanids are the temnodontosaurs, another group that likely originated in the Late Triassic but had their heyday in the Early Jurassic. *Temnodontosaurus* is among the first ichthyosaurs to become known, its scientific debut happening in 1814. The original specimen – a skull more than 1 m (3¼ ft) long, associated with a skeleton 5 m (17 ft) long – was discovered by the famous fossil-finding family of Dorset, England, the Annings, and described by Everard Home. William Conybeare, Richard Owen and others identified it as a species of *Ichthyosaurus* called *I. platyodon*. Not until 1889 did English naturalist Richard Lydekker realise it was worthy of its own name: he called it *Temnodontosaurus*, meaning cutting-tooth lizard, after the two keels on its teeth. Some temnodontosaur teeth have three keels and are triangular in cross-section, but most have none.

By the late 1800s, additional *Temnodontosaurus* species were known from the Holzmaden region, northern France, and Whitby in England. They show that temnodontosaurs were large, typically around 6 m (19¾ ft) long, sometimes 9 m (29½ ft) long or more, and sometimes with skulls more than 2 m (6½ ft) long. Possible temnodontosaur fragments from the Lias of Dorset suggest greater lengths of around 15 m (49 ft).

Complete skeletons – mostly from Germany – show that temnodontosaurs were long-bodied, long-tailed and long-limbed relative to thunnosaurs such as *Ichthyosaurus*. Their limbs were long and tapering, and both their fore- and hindlimbs were more similar in size and shape than they are in thunnosaurs. A vertical tail fin was present, but it was less crescentic than that of thunnosaurs. Temnodontosaur limbs are composed of block-like bones and include three, perhaps four, digits. Notches are present on the leading edges of some of the forelimb bones. Combined, these features give temnodontosaurs a more archaic body plan than thunnosaurs, and they were probably slower and less agile.

In life, the size of the tail of *Suevoleviathan* must have been among its most notable features. The possibility exists that the tail was boldly patterned or brightly coloured and used in sending signals.

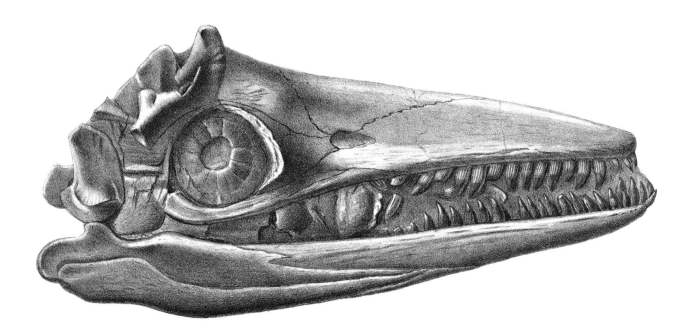

The heavily built skull of *Temnodontosaurus eurycephalus*. A mass of bone at the back of its jaws has been identified as belonging to another ichthyosaur species. However, it is more likely from the *T. eurycephalus* itself.

Temnodontosaurs vary in snout and jaw shape. *T. platyodon* has long, robust jaws. Others (such as *T. acutirostris* from Yorkshire) have slender jaws that taper to a point. *T. azerguensis* from France has exceptionally slim jaws and a reduced, probably absent, dentition. At the other extreme is *T. eurycephalus*, known from a skull discovered at Lyme Regis some time prior to 1881. It has short, deep jaws and a relatively low number of short, heavily sculpted, keeled teeth and looks vaguely like a killer whale skull.

The diversity observed across these species is substantial, so much so that classifying them within the same genus is beyond the variation ordinarily accepted for an ichthyosaur genus. It's conceivable that they will prove to occupy different positions on the family tree, and thus require different generic names.

The giant size, robust jaws and massive, keeled teeth of some temnodontosaurs show that they were predators of large vertebrates such as smaller ichthyosaurs and plesiosaurs, and also cephalopods. This is confirmed by stomach contents known for *T. platyodon* and the similar *T. trigonodon*. An analysis of jaw muscles in *T. platyodon* found evidence for a bite about twice as powerful as that of a large living crocodile, and in keeping with an ability to kill large vertebrates. *T. azerguensis* was built for a different diet, since its slender, toothless jaws and reduced jaw muscles look suited for small, soft prey.

Nothing is known about temnodontosaur social behaviour. Several specimens preserve broken and healed ribs, wounds on the face and snout, and damaged forelimbs and pectoral girdles. The causes of these injuries are impossible to determine, but some might be the result of fights with other members of their species.

The slender skull of *Temnodontosaurus azerguensis*, preserved lying on its upper surface. Like most temnodontosaurs, it was a big animal, with a skull 1.7 m (5½ ft) long and a total length exceeding 10 m (30 ft).

A final point worth noting is that temnodontosaurs are strangely restricted in distribution. Giant, marine animals have the capacity to cross oceans and be globally distributed, yet virtually all temnodontosaurs are from western Europe, and thus from the Western Tethys Ocean. Within this region, temnodontosaur fossils come from sediments laid down in temperate, subtropical and tropical waters, so it appears that they inhabited diverse environments. It is possible that they were limited in distribution, but far more likely is that their distributions were extensive, and that we simply lack the fossils needed to prove it. The only clue so far that this speculation has merit is a chunk of jaw, 60 cm (23½ in) long, from the Lower Jurassic of the Atacama Desert in Chile. This one fossil indicates that temnodontosaurs occurred throughout Tethys Ocean, and throughout Panthalassa, too.

This large, famous skull – discovered by the Annings in 1811 – belongs to *Temnodontosaurus platyodon* and shows the long, robust jaws typical of this species. This specimen is today on display at the Natural History Museum, London.

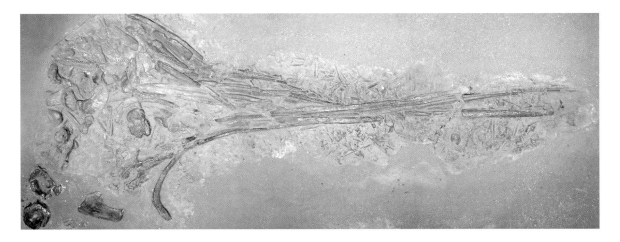

LEPTONECTIDS

The Western Tethys Ocean was home to another parvipelvian group during Early Jurassic times, the leptonectids, also known as the eurhinosaurs. In several respects, leptonectids are similar to temnodontosaurs. Their limbs are long, slender and formed of three or four digits, the fore- and hindlimbs are similar in size, their bodies and tails are long relative to those of thunnosaurs, and the tail bend is less extreme than it is in thunnosaurs. A few features make leptonectids distinct, including an upper arm bone with a strongly constricted shaft, a shortened, reduced rear to the skull, and small, slender teeth with smooth surfaces.

The most prominent leptonectid feature is a slender snout and lower jaw, both of which are shallower than the tall, rounded cranium. The eyes are directed forwards more than is usually the case in ichthyosaurs. The result is a swordfish-like look. This is especially obvious in *Eurhinosaurus* from Yorkshire, Holzmaden, Luxembourg, France and Switzerland. This was a large animal, as long as 7 m (23 ft), where the shortened lower jaw results in a hugely accentuated overbite. An overbite is also present in *Excalibosaurus* from England – named with reference to King Arthur's mythical sword Excalibur – a geologically older, smaller, more slender leptonectid. There is also an overbite in the English *Wahlisaurus*.

The majority of leptonectids belong within *Leptonectes*, species of which are known from the Early Jurassic of England, Germany, Belgium, Spain and Switzerland. These are variable in size and snout length, some being long-snouted giants (*L. solei* was around 7 m, 23 ft long), others being short-snouted dwarfs (*L. moorei* was only around 2 m, 6½ ft long). Some *Leptonectes* specimens come from the Upper Triassic, one of several pieces of evidence proving that the parvipelvian radiation was underway before the

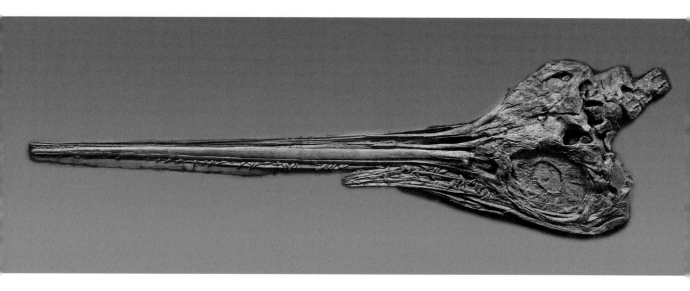

The skull of *Eurhinosaurus*, famous for its prominent overbite and very slender jaws. *Eurhinosaurus* has been known to science since the 1850s and was initially regarded as a species of *Ichthyosaurus*. While superficially swordfish-like, it also recalls certain fossil whales in snout and jaw shape.

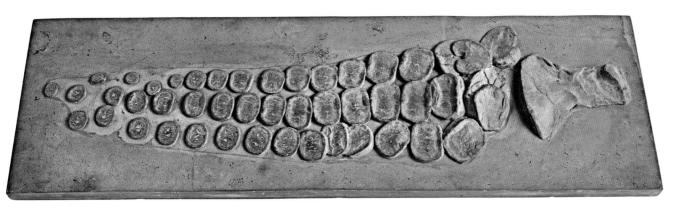

This leptonectid limb displays details typical of the group: the number of digits is low, the limb is long and slender, and the majority of the bones have rounded edges. The tips of ichthyosaur limbs are often incomplete since the small bones that formed them tended to be lost during decomposition.

Jurassic. As with the *Temnodontosaurus* species, the *Leptonectes* species are so variable that doubt exists as to whether they should be included within the same genus, and some studies find them to be well separated on the family tree. If such conclusions are supported by further work, new names will be needed for some species.

Views on how leptonectids might have lived have varied. Their slim jaws look suited for small prey, and their temporal openings are also small, meaning they had small jaw-closing muscles and a weak bite. The swordfish-like form of *Eurhinosaurus* has led to suggestions that it behaved in swordfish-like fashion, swimming at speed in pursuit of pelagic prey, which it injured with swipes of its snout. According to this view, leptonectids were fast and hunted in open waters. *Eurhinosaurus* is, however, different from swordfish and their kin in that its upper jaw is toothed for its length, whereas the swordfish rostrum is toothless. The swordfish rostrum also has a flattened cross-section and sharp cutting edges, which *Eurhinosaurus* lacks.

A different suggestion – inspired by the small, delicate teeth, huge eyes and enormous limbs of these ichthyosaurs – is that they were slow-moving foragers that hunted for small animals close to the seafloor. Our knowledge of what leptonectids were like and how they lived is in its infancy, and more study is needed.

Temnodontosaurs and leptonectids are both long-limbed, long-tailed and, mostly, long-snouted. Most studies find them to be closely related but on distinct branches of the parvipelvian tree, outside Thunnosauria but close to it. Temnodontosaurs and leptonectids are, however, sufficiently similar that they might represent a clade. *Suevoleviathan* is also temnodontosaur-like in some aspects of skull anatomy, so the possibility exists that it is also part of a temnodontosaur-leptonectid clade.

ICHTHYOSAURIDS

A fourth group evolved at about the same time as the suevoleviathanids, temnodontosaurs and leptonectids during the Late Triassic. This time it contained species familiar to non-specialists. This group is Thunnosauria,

This *Ichthyosaurus* skull preserves a near-complete dentition. The several *Ichthyosaurus* species can be distinguished thanks to the shape and proportions of their skull bones, as well as by numerous other details present throughout the skeleton.

Ichthyosaurus specimens are mostly preserved on their sides, though a few are preserved on their backs or bellies. Complete, articulated skeletons like this one were probably buried quickly, in cases because they sank into soft sediments. When this happened, at least some of their original organic tissue (like skin and muscle) was preserved in addition to the bones.

its name meaning 'tuna lizards', a reference to their tuna-like shape. Part of the thunnosaur story involves the evolution of an increasingly streamlined, efficient shape and an ability to travel fast across distance. Several earlier ichthyosaur groups had already crossed the great oceans of the time, but thunnosaurs probably did this regularly, many species foraging and hunting hundreds of kilometres from land, out in the vastness of Tethys and Panthalassa.

The oldest thunnosaurs belong to Ichthyosauridae, and among this group is *Ichthyosaurus communis* from the Early Jurassic of southern England, the archetypal ichthyosaur. *Ichthyosaurus communis* is one of several *Ichthyosaurus* species. Hundreds (perhaps thousands) of specimens are known, the majority of which are from Street in Somerset and Lyme Regis and Charmouth in Dorset. A few *Ichthyosaurus* specimens are known from Portugal, Belgium, Germany, Switzerland and Canada. *Ichthyosaurus* is long-jawed and mid-sized, mostly between 2 and 3 m (6½ and 10 ft long). It has large eyes, conical, robust, striated teeth that line the jaws all the way back to the eyes, and a robust cranial region.

The upper arm bone in *Ichthyosaurus* indicates that the forelimb's base was muscular and important in braking, turning and accelerating. In contrast to temnodontosaurs and leptonectids, the ichthyosaurid forelimb always has at least four parallel digits formed of tightly fitting, closely packed phalanges that are arranged like rectangular tiles. There are sometimes more than 25 phalanges in the longest digits. The digits sometimes bifurcate along their length and there are usually one or two extra digits parallel to the limb's

Malawania from the Early Cretaceous of Iraq was a 'living fossil' of its time – a late-surviving relict most similar to ichthyosaurids from the Early Jurassic, more than 50 million years earlier. Its name is derived from the Kurdish word for 'swimmer'.

trailing margin. An extra digit on the limb's leading margin is sometimes present, too. One result of this proliferation of digits and phalanges is that the end of the limb sometimes has more digits than the part close to the upper arm, as many as nine. In fact, *Ichthyosaurus* is variable in limb structure, animals from different locations varying in digit number, phalangeal number, how much bifurcations they have, and how much notching they have along the limb's leading margin.

This variation is linked to the question of how many *Ichthyosaurus* species there are, since individuals also vary in size, snout length, eye size, the configuration of the skull bones and much else. Recent studies, mostly led by ichthyosaurid specialist Dean Lomax, indicate that there are around eight species, ranging from the small, short-snouted *I. breviceps* to the larger, long-snouted *I. somersetensis*. These species were mostly around 2 m (3 1/4 ft) long, the largest reaching 3.5 m (11½ ft). Species of another ichthyosaurid – *Protoichthyosaurus* – are known from the Early Jurassic of southern England and Wales.

It was long thought that ichthyosaurids had disappeared by the Middle Jurassic. However, a fossil from Kurdistan in Iraq challenges that idea. This is *Malawania*, discovered in 1952 but not named until 2013. *Malawania* resembles *Ichthyosaurus* in forelimb structure, but has a shorter, broader humerus with a unique triangular prominence at its upper end. The most significant thing about *Malawania* is that it is from the Early Cretaceous. This means that ichthyosaurids survived the Jurassic-Cretaceous extinction event, albeit at low diversity and in just one location.

STENOPTERYGIIDS

Ichthyosaurids are abundant in the Lower Jurassic rocks of England, but remain rare in France, Germany and elsewhere. These countries have, however, yielded numerous specimens of a similar but geologically younger Early Jurassic group, the stenopterygiids, named after *Stenopterygius* from the Posidonia Shale.

Several *Stenopterygius* species have been named, differing in vertebral count, the proportional size of the skull, length of the snout, forelimb length, and their tendency towards tooth loss. The large, short-snouted *S. uniter* and mid-sized, long-snouted *S. quadriscissus* underwent tooth reduction as they matured, adults of *S. quadriscissus* usually being toothless. Other species exhibited no such tendency. The species vary in size, from 2 to more than 3.5 m (6½ to 11½ ft). This variation means that doubt exists as to whether these animals should be classified together, and they rarely group together in phylogenetic studies. Another stenopterygiid – *Hauffiopteryx* – is known from Germany while *Chacaicosaurus* from Argentina might also belong to this group.

S. quadriscissus is the best known stenopterygiid. It is also among the most important of ichthyosaurs in terms of what it tells us about their biology. Many Posidonia Shale specimens died while pregnant, and so many were carrying unborn young that the Holzmaden area was likely a birthing site. Perhaps pregnant females made a migration to give birth here because it provided abundant feeding or shelter for their newborns.

Stenopterygius also plays an important role in our understanding of ichthyosaur evolution since its anatomy and proportions appear intermediate between those of ichthyosaurids and the even more streamlined ophthalmosaurids. Stenopterygiids have especially small hindlimbs which are about half the size of the forelimbs, and a pelvis where the pubis and ischium bones are fused together. They are ophthalmosaurid-like in these respects. Stenopterygiids also share with ophthalmosaurids a prominent, square projection on the leading edge of the scapula, called the acromion. This explains why stenopterygiids and ophthalmosaurids are united within the clade Baracromia, meaning heavy acromion.

This extremely well-preserved *Stenopterygius* specimen from Germany preserves a second, small ichthyosaur within its body cavity. Perhaps this is an unborn baby, or the remains of a meal. Fish bones are also preserved in the animal's stomach region.

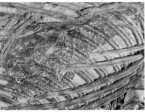

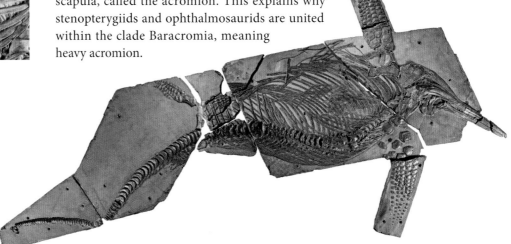

OPHTHALMOSAURINES

During the early part of the Jurassic, around 190 million years ago, a lineage of *Stenopterygius*-like ichthyosaurs evolved broad, thickened forelimbs and a more teardrop-shaped body. Their descendants were the ophthalmosaurids, a long-lived group that were the final flowering of ichthyosaurs and the most species-rich ichthyosaur group in history. Many features that make ophthalmosaurids different from other thunnosaurs and parvipelvians are related to increased streamlining. The anatomy of the back of the skull shows that they had a reduced range of head motion relative to their ancestors, perhaps to keep the skull stable during swimming. Ophthalmosaurid innovations include an increase in eye size and nostril complexity, and high variation in forelimb anatomy and tooth shape. Most species were 3 to 6 m (10 to 19¾ ft) long.

Ophthalmosaurids include two major groups: ophthalmosaurines and platypterygiines. Ophthalmosaurines have small teeth and a skull where the eyes are so large that they make the cranium much taller than the snout. Platypterygiines have enlarged, robust teeth with quadrangular roots, a skull where the difference in height between the cranium and snout is not so obvious, a strongly muscled upper arm and extra digits.

Artwork and diagrams often create the impression that ophthalmosaurid limbs were narrow and blade-like. They were actually thick and chunky. Massive muscle sites in the upper arm show that the forelimb's base was extremely muscular and most of the bones forming the limb were thick and built like hockey pucks. Ichthyosaur limbs in general are unusual, with their numerous rows of tightly packed, tile-like phalanges, extra bones at the tips of their digits and even entirely new digits. Platypterygiines took this to an extreme, some species possessing 10 digits in each forelimb, the longest with more than 30 phalanges. Presumably, these wide, lengthened limbs were important in steering and manoeuvring, or in generating lift.

Ophthalmosaurines are a mostly Jurassic group, though two or three lineages survived into the Early Cretaceous. Their best-known member is

Stenopterygius was sleek and superficially dolphin-like. Our knowledge of ichthyosaur pigmentation remains sparse. Complex, dolphin-like patterns like those shown here are plausible, but so are plainer looks where the animal was mostly black or grey.

Ophthalmosaurus, described from the Oxford Clay in 1874. Many ideas about ichthyosaur vision, diving abilities and ecology are based on this animal. Several *Ophthalmosaurus* specimens lack teeth, so a view once favoured by experts was that *Ophthalmosaurus* was toothless, perhaps because it was a specialized squid predator. Toothlessness makes sense for squid eaters, in part because teeth act as sites which a squid will grab to prevent its being swallowed by the predator. It turns out that numerous small, slender, conical teeth lined the jaws, their loose attachment meaning that they often dropped out after death. These teeth indicate that *Ophthalmosaurus* was a predator of small, soft prey, these probably including fish as well as squid.

The eyeballs of *Ophthalmosaurus* could be 22 cm (8¾ in) wide, making them the largest, relative to body size, of any ichthyosaur. Thanks to complete sclerotic rings (the bony structures embedded within the eyeballs of most vertebrate animals), we can also work out pupil size, since this corresponds to the size of the aperture in the middle of the ring. In 1999, Ryosuke Motani and colleagues showed how pupil size in ichthyosaurs can be linked to their ability to see in the dark. They concluded that *Ophthalmosaurus* could detect light around 500 m (1,640 ft) down, in the mesopelagic or twilight zone. A view popular among ichthyosaur researchers is that *Ophthalmosaurus* was a specialist deep-diver, routinely foraging in dark waters off the continental shelf. This is possible, and it could be that these animals were able to dive much deeper than 500 m (1,640 ft). However, *Ophthalmosaurus* fossils mostly come from continental seas where the water was not especially deep, and it might be that they made a living in far shallower seas. Maybe they were adaptable and did both.

Ophthalmosaurus was widespread. Fossils have been reported from France, Russia, the USA, Argentina and elsewhere. However, some of these probably represent different animals. Jurassic fossils from the North American Sundance Sea, for example, are sometimes labelled *Ophthalmosaurus* but might warrant their own genus: *Baptanodon*. Numerous other ophthalmosaurines are known, with both European Russia and Spitsbergen being important for the group. Species of the small *Nannopterygius* and larger *Arthropterygius* were present in the Western Tethys Ocean and Boreal Ocean, and both seem to have moved between these areas during their history. Our view of ophthalmosaurine diversity is currently in a dynamic phase, with new, sometimes conflicting studies and conclusions appearing regularly.

PLATYPTERYGIINES

The platypterygiines (termed undorosaurids or brachypterygiids by some authors) are an even more complex group. Our knowledge of these animals began with the 1907 naming of *Platypterygius* from the Early Cretaceous of Germany. The original specimen (an articulated partial skeleton) was destroyed during World War II but other Early Cretaceous European fossils represent members of the same genus. *Platypterygius* was around

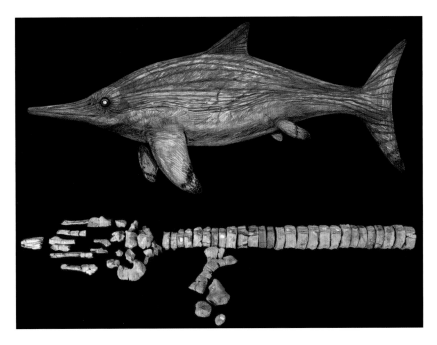

Acamptonectes from the Early Cretaceous of western Europe, named in 2012, is one of several recently named ophthalmosaurines. It was slender-snouted and similar to *Ophthalmosaurus* in many respects but possessed an unusual stiffened vertebral column.

5 m (16½ ft) long and had stout, conical teeth and a broad, deep base to its long snout.

In the years following 1907, other Cretaceous platypterygiines were discovered in Russia, the USA, Canada, Australia, Argentina, Colombia and elsewhere. Opinions differed on how to classify them. Some were given their own genera while others were regarded as new species of *Platypterygius*. By the early 2000s, what might be considered the 'mainstream' position is that they should all be regarded as variants of *Platypterygius*, a view which requires the persistence of this genus for an exceptional 35 million years. This would make Cretaceous ichthyosaur history a non-event marked by stasis and monotony. Things began to change in 2010, both as new Cretaceous platypterygiines were discovered (in particular *Athabascasaurus* from Canada), and as European and Russian specimens were re-analyzed. As argued by Valentin Fischer and colleagues in a series of studies, the animals lumped together in *Platypterygius* exhibit more diversity in snout, jaw and tooth shape than previously appreciated. Furthermore, they do not group together on the same part of the family tree. The name *Platypterygius* should be restricted to certain western European animals, and the others warrant different names. New discoveries show that Cretaceous platypterygiines possessed a variety of body sizes and feeding adaptations. They represent at least 10 genera, with more awaiting publication at the time of writing.

Several of these have broad, heavily built skulls and large, often heavily worn teeth, features that show they were predators of large invertebrates as well as fishes, birds and other marine reptiles. Stomach contents from an Australian platypterygiine confirm this, since it had eaten birds and

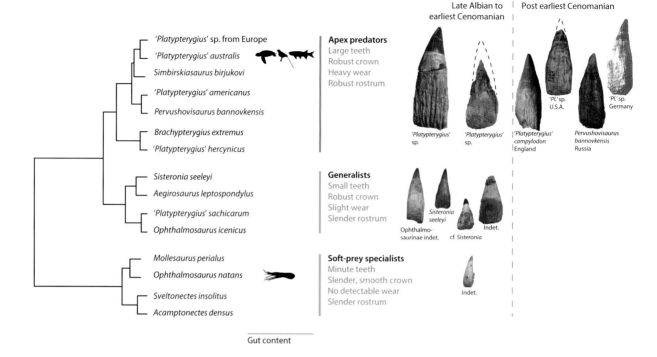

This tree-like diagram shows how ophthalmosaurines and platypterygiines group together on the basis of anatomy and feeding behaviour. Platypterygiines – equipped with massive teeth and able to consume bony prey including turtles and birds – survived later into the Cretaceous than ophthalmosaurines.

turtles. In *Kyhytysuka* from Colombia, the teeth show heterodonty, meaning difference in shape, and group into five sections along the jaws, some suited for piercing, some for cutting and some for crushing. The jaws of this animal could be opened to a considerable angle of 75 degrees and were built to withstand the transfer of considerable force. This is consistent with the view that certain platypterygiines were top predators, able to kill, dismember and swallow large prey, probably including other marine reptiles.

Nostril shape is unusual in some platypterygiines. Thunnosaurs in general have an oval nostril but ophthalmosaurids have a bony flange that descends from the nostril's upper margin such that the nostril is constricted halfway along its length. In the small platypterygiine *Sveltonectes* from Russia, a vertical bar divides the nostril in two. In later forms, this is a thick bony pillar, separating the nostril into a small, horizontal opening at the front and a larger, deeper opening closer to the eye. Why these ichthyosaurs evolved these two-part nostrils is unclear. One suggestion relates to the presence of the salt-excreting glands these animals likely possessed near their eyes. In order that the salt solution excreted by these glands is removed by water, animals benefit from structures that slow or divert water around the duct where the solution leaves the head. Maybe the bony pillar served this role, since it presumably slowed and diverted water into the larger, more vertical nostril opening, the place where the salt gland was likely sited.

Cretaceous platypterygiines, then, were hardly static or monotonous, but diverse and with a complex history. At least eight platypterygiine lineages survived across the Jurassic–Cretaceous boundary. Indeed, we know that

platypterygiines were diverse in the Jurassic, too. Among the most interesting of these is *Grendelius*, named from the Kimmeridge Clay by ichthyosaur expert Chris McGowan in 1976. Its name refers to Grendel, the monster of the first century Anglo-Saxon poem *Beowolf*. Several platypterygiines from Russia originally named *Otschevia* appear to be additional species of *Grendelius*. The jaw tips in these animals are blunt and rounded, not slim and pointed as they usually are in ophthalmosaurids. The base of the snout is not that different in height from the top of the skull, and the region behind the eyes is longer and more massive than is typical. The *Grendelius* skull is also proportionally large, at about 75% the length of the ribcage – in other ophthalmosaurids, the skull is usually less than 60% the length of the ribcage. These animals look well equipped to grab and subdue large prey.

For all our knowledge of ichthyosaur diversity, we have yet to discover any species that inhabited estuaries or freshwater. It might be that ichthyosaurs never invaded these environments, perhaps because other reptile groups prevented this, or because there was never any evolutionary pressure for them to do so. It might be, however, that estuarine or freshwater ichthyosaurs did exist but haven't yet been properly recognized. Hints that they did exist come from rare, fragmentary platypterygiines from the Lower Cretaceous freshwater sediments of southern England, specifically the Purbeck Limestone in Dorset and the Wealden in West Sussex.

Platypterygiines include the last of the ichthyosaurs. Fossils show that as many as seven platypterygiines were in existence around 114 million years ago, but that they declined late in the Cenomanian age around 94 million years ago and failed to survive beyond it.

> **This spectacular *Grendelius* skull was discovered in Norfolk, England, in 1958. It now seems that several Russian ophthalmosaurids – originally named *Otschevia* – represent additional *Grendelius* specimens, in which case *Grendelius* is better known than originally thought.**

6 | LONG NECKS, BIG MOUTHS:
THE PLESIOSAURS

THE FOUR-FLIPPERED, often long-necked plesiosaurs exceed ichthyosaurs in being the most familiar, most famous Mesozoic marine reptiles of all. English palaeontologist William Buckland described them as resembling "a serpent threaded through the body of a turtle" during the 1830s, only a few years after plesiosaurs became known. When Richard Owen published his review of British fossil reptiles in 1841, it was known that plesiosaurs included short-necked, big-headed kinds with enormous teeth, such as *Pliosaurus*. Discoveries made in the Upper Cretaceous of the USA revealed the existence of the giant, extremely long-necked elasmosaurids, and so remarkable was their appearance that they soon became the sort of plesiosaurs depicted most frequently in artwork and books.

Part of our familiarity with plesiosaurs comes from the idea, prevalent in popular culture, that plesiosaurs might have survived to the present. This notion was first outed in the 1840s to explain apparent sightings of sea monsters, and was repeatedly endorsed by later writers seeking to make sensational claims. When Loch Ness in Scotland was said to be the haunt of a monster in the 1930s, this too was framed as a 'living plesiosaur'. Such was – and is – the public's adoration of Nessie that the terms Loch Ness Monster and plesiosaur became almost synonymous. In reality, the sea and lake monster sightings that people have in mind when discussing 'living plesiosaurs' describe objects that have no similarity to plesiosaurs at all. They

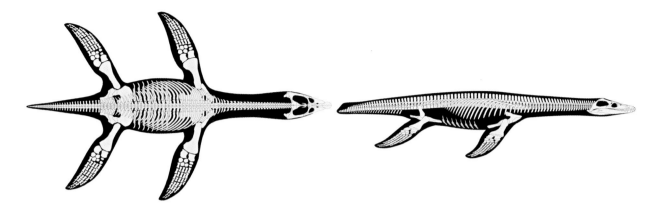

All plesiosaurs (this is the Early Jurassic *Rhomaleosaurus*) share the same four-flippered, short-tailed body plan. The flippers were wing-like in shape and function. Neck length and head size were highly variable within the group.

include turtles or large fish entangled in fishing gear, and misidentified seals, whales and swimming deer. Nor is there reason to think that plesiosaurs survived beyond the Cretaceous. Despite this, and much to the chagrin of palaeontologists, 'Loch Ness Monster' is still the public's primary point of reference when hearing the word 'plesiosaur'.

To return to real plesiosaurs, this was an extremely successful, globally distributed group that originated in the Triassic and then thrived worldwide throughout the Jurassic and Cretaceous, a span of over 130 million years. In a sense, plesiosaurs were conservative, since all species possessed the same body plan. This involved a stiff, compact body where the enlarged, plate-like bones of the pectoral and pelvic girdles were arranged on the underside, the limbs were tapering flippers devoid of claws, and the tail was short and robust. The interlocking rib-like gastralia formed a basket-like arrangement on the underside. All were predatory, their diets ranging from crustaceans, fishes and soft-bodied invertebrates to other marine reptiles and maybe even terrestrial animals such as dinosaurs. It can also be said that plesiosaurs were diverse, varying substantially in proportions, ecology and lifestyle.

PLESIOSAUR EVOLUTION AND CLASSIFICATION

Plesiosaurs are variable in skull shape, size, neck length, and in the shapes of their ribs, limb girdles and limbs. They have traditionally been sorted into two groups – the long-necked plesiosauroids and the short-necked pliosauroids – a division created by DMS Watson in 1924. This belies, however, a complex nineteenth- and twentieth-century history in which the vertebrae, ribs and pectoral girdle were used to reconstruct plesiosaur evolution. In 1940, Theodore White argued on the basis of diversity in the pectoral girdle that plesiosaurs could be grouped into eight families, and that long and short necks evolved several times independently. Studies by Samuel Welles, published during the 1940s to 1960s, reversed White's conclusions and reinforced the plesiosauroid versus pliosauroid division. This simple view of plesiosaur classification and evolution became the 'textbook' view.

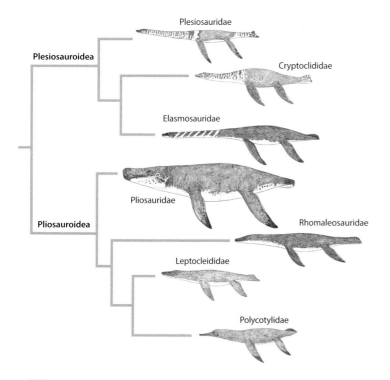

This diagram depicts 'traditional' thinking on the shape of the plesiosaur family tree. It was thought that plesiosaurs split early in their evolution into the long-necked group Plesiosauroidea and the short-necked group Pliosauroidea. More recent studies reveal a more complex picture.

A new era in our understanding of plesiosaur evolution was initiated in 1993 by Robert Bakker, the American palaeontologist best known for his work on dinosaurs. Bakker argued that plesiosaur evolution had been misunderstood and that the group's history was dynamic. He suggested that it involved several extinction events and the repeated evolution of similar body plans. The animals grouped together as pliosauroids were not members of a single lineage, Bakker argued, but had emerged more than once from long-necked ancestors. He concluded that the long-necked, multi-toothed cryptoclidids were of mysterious ancestry while elasmosaurids were closer to the mostly short-necked leptocleidids and polycotylids, previously regarded as pliosauroids.

Since 2001, studies by Robin O'Keefe, Hilary Ketchum, Roger Benson, Patrick S Druckenmiller and others have examined plesiosaur evolution anew. They have mostly confirmed Bakker's proposals. Because the terms 'plesiosauroid' and 'pliosauroid' as traditionally used refer to repeatedly evolved body shapes, not to clades, experts have taken to using 'plesiosauromorph' and 'pliosauromorph' to describe these two forms. It should be noted, however, that the technical term Plesiosauroidea is still in use for the clade that includes *Plesiosaurus* and kin, and that Pliosauroidea is still in use for the clade that includes *Pliosaurus* and kin.

The mostly Early Jurassic rhomaleosaurids – traditionally regarded as pliosauroids – might be one of the first plesiosaur groups to evolve. Leptocleidids, polycotylids and elasmosaurids are united within Xenopsaria, a name that means 'strange fish-catchers'. The idea that elasmosaurids are close cousins of the pliosauromorph leptocleidids and polycotylids seems radical in view of the traditional classification system, but studies published throughout the 1990s and twenty-first century have repeatedly emphasized their closeness.

Several of the points here are the topic of debate, and future work and discoveries may result in new proposals. One area of disagreement concerns xenopsarian history. In a 2013 study, Benson and Druckenmiller argued that xenopsarians originated at the start of the Cretaceous, exploding onto the scene late in plesiosaur history. Other experts, however, think that certain Jurassic fossils are pre-Cretaceous members of the leptocleidid or polycotylid lineages. If this is true, xenopsarian evolution was a slower, more drawn-out event. Regardless of debates such as this, what this new, complex view of plesiosaur evolution means is that giant size – anything exceeding 6 m (19¾ ft) – evolved at least three times, that especially long necks evolved at least three times, and that the pliosauromorph shape also evolved at least three times.

A final complication concerns the word 'plesiosaur' itself. Because this has sometimes been used specifically for plesiosauromorphs, some experts argue that things are made confusing when it is used for the group that includes all members of Plesiosauria. They therefore recommend use of the term 'plesiosaurian' when referring to Plesiosauria as a whole. However, the term 'plesiosaur' is so familiar in ordinary language that changing to a technical, obtuse alternative is not sensible. For that reason, this proposal is ignored here.

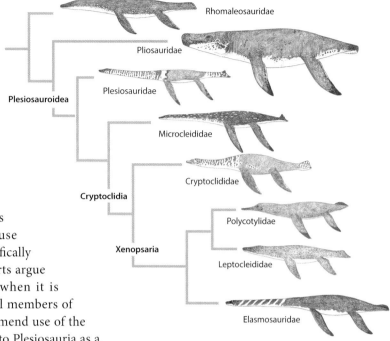

Plesiosaur evolution was complicated. The relatively short-necked rhomaleosaurids and pliosaurids might be among the first plesiosaur groups to have evolved, but early members of both groups possessed relatively long necks. 'Pliosauromorph' groups like polycotylids evolved later on too, this time from different ancestors.

ANATOMY

Plesiosaurs are remarkable for a list of reasons, and not just for having long necks. Their pectoral and pelvic girdles are among the most unusual ever evolved, and the robust gastral basket and short, stout tail are also unusual.

Plesiosaur limbs are wing-shaped flippers. Mobile elbow, wrist, knee and ankle joints are absent, and the only areas where muscular movement was possible were the shoulder and hip. The bones closest to the body – the humerus in the fore flipper and femur in the hind – are similar, so much so that experts struggle to tell them apart when discovered in isolation. For this reason, they are often referred to collectively as propodials. A propodial has a rounded head, cylindrical shaft, and flattened, broadened end. Rough, raised areas on the shaft mark where the muscles powering the limbs were attached. The other large bones of the limbs – the radius and ulna in the fore flipper and tibia and fibula in the hind – are rectangular or rounded, flattened bones, very different from the long, cylindrical bones of terrestrial reptiles. Collectively, these bones are the epipodials, and it is again difficult to work out whether they are from the fore or hind flippers when found in isolation. In some plesiosaur groups, the epipodials are polygonal and fit together tightly, both to each other and to the surrounding bones.

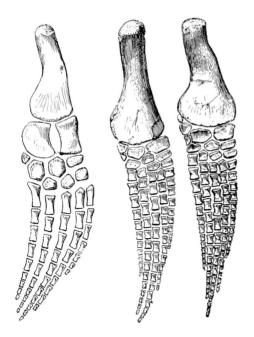

Plesiosaur limbs were highly specialized, wing-like flippers. Over the course of evolution, the forelimbs and hindlimbs became increasingly similar, and there is little doubt that both limb pairs served the same function. The image at far right is a forelimb but the other two are both hindlimbs.

Soft tissue outlines show that plesiosaur flippers (one is shown at far right) were similar to those of sea turtles, penguins and sea lions (shown here from left to right) in being wing-shaped and with a flexible trailing margin.

Beyond the epipodials, three rows of rounded, polygonal and rectangular bones represent modified wrist and ankle bones as well as those that ancestrally formed the main sections of the hand and foot. Additional bones unique to plesiosaurs are sometimes present adjacent to, or among, these bones. Overall, this configuration provided a stiffened, immobile, broad middle section to the flipper, built to conduct muscular force to the flipper's end part without deforming. Finally, the remainder of the flipper – the long, tapering end section – is supported by the numerous hourglass-shaped bones of the fingers and toes, termed phalanges. Reptiles ancestrally have between two and five phalanges, but plesiosaurs evolved extra ones such that their longest digits have more than 10. Plesiosaur digits are pressed together tightly and are sometimes connected by interlocking joints, so there wasn't any possibility of the digits spreading apart.

How much soft tissue was present around the flipper's base is uncertain and it might be that plesiosaur groups differed in this detail. The long, narrow propodials of some cryptoclidids and polycotylids suggest that the basal section of the flipper was much narrower than the flipper's middle. In contrast, the broader, shorter propodials of some other groups perhaps meant that the flipper's base was deeply submerged in fat and skin. As for the rest of the flipper, it is suspected (based on comparison with the flippers of living animals) that a flexible margin extended beyond the edge of the limb skeleton on the trailing side. The presence of such a structure is confirmed by a few specimens. The same specimens show that the flipper tips were strongly curved in some plesiosaur groups, a soft tissue extension continuing beyond the digital tips.

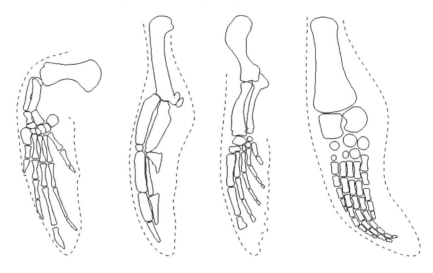

SWIMMING

The unusual limb girdle arrangement, compact body and wing-like limbs of plesiosaurs have inspired many scientists to wonder how these creatures swam. Their limbs were clearly used for underwater propulsive – in rowing, sculling or flapping – and their enlarged limb girdles look like muscle attachment sites. But how exactly did they move? This question has been difficult to answer because plesiosaurs are very different from all living animals. As a result, experts have disagreed on which swimming style was most likely. Probably the oldest suggestion (it was first put forward by William Conybeare in 1824) is that plesiosaurs used a flying motion, meaning both limb pairs were flapped vertically, with the downstroke and upstroke being near-symmetrical. In 1924, London-based zoologist David M S Watson proposed a second model. Watson found that those muscles anchored to the humerus and femur were arranged to pull the limbs forwards and backwards, rather than up and down. He therefore argued that plesiosaurs used a rowing action, the thrust being generated as the limbs were pulled backwards. The limbs would then be rotated by 90 degrees and 'feathered' (meaning made parallel to the direction of movement) to minimize resistance on the recovery stroke, similar to the oar of a rowing boat. A problem with this model is that it fails to produce thrust or lift on the recovery stroke. Watson's rowing model was widely accepted for the next several decades.

A renewed interest in plesiosaur swimming emerged in the 1970s and 1980s as specialists began to look anew at this subject. In a 1975 study, plesiosaur expert Jane A Robinson revived the flying theory, concluding that the shape of plesiosaur limbs and their ranges of movement suggested they swam in a way similar to penguins and sea turtles. Other experts took these ideas further. One feature that makes plesiosaurs different from modern swimmers is that both limb pairs are wing-like, suggesting that both had the same function. Why have two pairs when one is good enough for other swimming groups?

Plesiosaurs are unique in possessing two pairs of wing-like flippers. Some experts have argued that they used their limb pairs in alternating fashion, the front pair undergoing downstroke while the hind pair underwent upstroke.

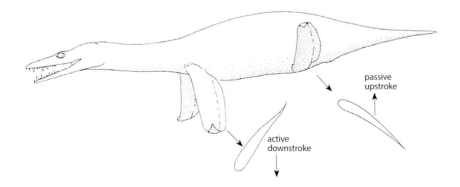

In 1982, German palaeontologists Eberhard Frey and Jürgen Riess argued that plesiosaurs had mastered a swimming style known as the alternating downstroke. They proposed that plesiosaur limbs generated thrust on the downstroke, but not on the upstroke. They suggested the two limb pairs moved in opposition, the front pair undergoing downstroke as the hind pair underwent upstroke, and vice versa. As a result, a swimming plesiosaur would generate thrust continuously. This model appears logical but fails to account for an important problem - vortices. A flapping wing or flipper produces a vortex on its leading edge every time it flaps up and down. These vortices detach at the top or bottom of the flapping stroke and are washed downstream. The alternating downstroke model requires that the vortices shed by the plesiosaur fore flipper during its downstroke are intercepted by the hind flipper on its upstroke, the result being untidy disruption of the vortices. If we assume that evolution favours efficiency, and we mostly do, this means that the alternating downstroke model can't work, and that another model must have existed, one where the hind flippers interacted efficiently with vortices produced by the fore flippers.

Much of the research discussed so far is theoretical. Such work has its place but is often unable to account for the complications that arise when actions are performed in real-life. Today, computer modelling, robots and an ability to analyze the positions and forces of objects in space using measuring equipment greatly improves our ability to understand animal locomotion.

A 2014 study led by aerospace engineer Luke Muscutt combined all these tools. Muscutt and his colleagues constructed a robotic rig fitted with accurate replica plesiosaur flippers and made it swim in a flume tank. Dyes were used to visualize the vortices, and sensors were used to measure the swimming forces produced by the flippers. By changing the timings of the fore and hind flipper motions, the robot was able to demonstrate that two flipper pairs could function more efficiently than a single pair. The model showed that precise timing of flipper movement allowed the hind flippers to avoid the vortices shed by the front pair but to capture some of their momentum, thus increasing efficiency. When this technique was used, the hind flippers increased their thrust by as much as 60 per cent and their efficiency by as much as 40 per cent.

At the moment, this research has only been performed using flippers based on those of an Early Jurassic rhomaleosaurid. There are plans, however, to test the performance of other plesiosaur groups, the members of which differed substantially in flipper shape.

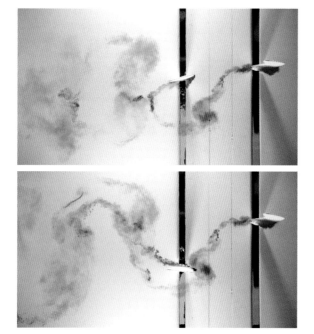

These images show how vortices – here made visible via the use of dyes (the red comes from the front flipper and the blue from the hind) – were shed from moving model plesiosaur flippers in a flume tank.

BEHAVIOUR AND BIOLOGY

The traditional view of plesiosaurs is that they were something like big reptilian seals, good at twisting and turning in pursuit of prey but otherwise adapted to throw their necks about in the air, and also to clamber on land. Numerous artistic reconstructions depict plesiosaurs behaving like this, but there was never a good reason for this view. Plesiosaurs lack features present in animals that move on land, such as flexible elbows and knees, and a bony connection between the hips and spine. They look fully adapted for aquatic life, and have necks, faces and eyes suited to moving, hunting and feeding in water, not air.

One piece of behaviour perhaps common to all plesiosaur groups is that they swallowed stones and kept them in the muscular part of the stomach. Once ingested, stones are termed gastroliths. Early plesiosaurs swallowed just a few small stones, effectively grit or gravel. But the swallowing of hundreds or even thousands of large ones – in some cases equivalent in size to a human fist and weighing as much as 1.4 kg (3 lbs) – was practised by elasmosaurids and leptocleidids. Some elasmosaurids are preserved with gastrolith masses of 13 kg (28¾ lbs). The gastroliths are not random stones, but are consistently dense and formed of materials such as quartz, chert, basalt or granite, which do not occur in the immediate area. The implication is that the animals travelled some distance to obtain them. Why plesiosaurs possessed gastroliths is disputed. Plesiosaur expert Mike Taylor has argued that gastroliths played a role in buoyancy and that plesiosaurs used them to control their position in the water column. Support for Taylor's argument comes from the presence of gastroliths in other aquatic vertebrates, and that those aquatic vertebrates that don't possess gastroliths use other methods of making their bodies heavy, such as developing thick, dense bone.

Masses of gastroliths have been discovered in the stomach regions of several plesiosaurs, most usually elasmosaurids. This collection weighed 8.8 kg (19½ lbs) and the largest stones were 11 cm (4¼ in) across.

An alternative view is that gastroliths help with digestion. Some gastroliths possess crescentic markings of the sort caused when such stones are ground together and have been discovered mixed up with the scales, teeth and bones of prey. This suggests they functioned in breaking up food. American marine reptile expert Mike Everhart has argued that even 13 kg of stones would have been insufficient to have any significant impact on the buoyancy of an animal that weighed several tonnes (as large plesiosaurs did), especially when its daily food budget would have involved the ingestion and excretion of greater amounts. Debate about the role and function of gastroliths continues, but a digestive role seems likely.

RHOMALEOSAURIDS

Among the most archaic of plesiosaurs are the rhomaleosaurids, a group of small, mid-sized and large animals mostly associated with the Early Jurassic of Europe. They persisted, however, to the end of the Middle Jurassic, likely had a near-global distribution, and might have included freshwater species. The group takes its name from *Rhomaleosaurus*, named in 1874 for a skeleton 7 m (23 ft) long discovered at Kettleness, Yorkshire, in 1848. This was originally described as a species of *Plesiosaurus*, but work by English palaeontologist Harry Seeley demonstrated it was something new. The name he chose means strong lizard, referring to its large size and robust skull. In 1853, this specimen was transferred to Dublin for display at a scientific meeting, but it later fell into neglect, the most extreme demonstration of which is that it was broken up by a sledgehammer so it could be transported more conveniently! Casts were made beforehand, however. One of the casts is at London's Natural History Museum, where its association with an information panel about Mary Anning makes many people think that *Rhomaleosaurus* – rather than *Plesiosaurus* – is Anning's original plesiosaur.

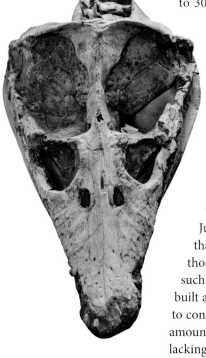

Seen from the front, the giant skull of *Rhomaleosaurus* (74 cm or 2½ ft long) reveals the broad-based, heavily built snout typical of large members of this group. Overlapping bony joints helped the skull resist deformation and break apart the bodies of large prey items.

Additional rhomaleosaurids have been named since, including other *Rhomaleosaurus* species. Several features unite these animals. The lower jaw bows outwards, the rear edge of the skull is slanted forwards, there tend to be deep grooves on the palate, and the pectoral girdle is of a sort where the scapulae are restricted to the edges. There are around 25 to 30 neck vertebrae. The rhomaleosaurid snout is never long, but the bones at the tip form a spoon-shaped extension separated from the rest of the face by a notch. *Rhomaleosaurus* is heavily built with a broad snout tip and subtle notch, but others – such as *Archaeonectrus* and *Macroplata* – have a longer snout and larger notch. These longer-snouted kinds were around 3–4 m (9¾–13 ft), smaller than *Rhomaleosaurus*, longer-necked and with a proportionally smaller skull. Rhomaleosaurid fore and hind flippers are about equal in length.

Several rhomaleosaurids, such as *Stratesaurus* from the earliest Jurassic, are small, lightly built and long-necked. They seem to have occupied several ecological niches, most of which involved the hunting of small fishes and crustaceans. The diversity of Early Jurassic rhomaleosaurids suggests that the group originated before that, in the Late Triassic, in which case members of the group older than those known await discovery. In contrast, the youngest rhomaleosaurids such as *Rhomaleosaurus* from the end of the Early Jurassic are big, heavily built and adapted for killing large prey. The *Rhomaleosaurus* skull was built to conduct force, withstand bending and twisting, and exert the maximum amount of damage to prey. Direct evidence for the diet of *Rhomaleosaurus* is lacking, but it likely hunted other marine reptiles and large fish.

A few fossils give indications of a poorly understood complexity to rhomaleosaurid history. While mostly associated with western Europe, *Borealonectes* is from the Middle Jurassic of the Canadian Arctic and *Maresaurus* from the Middle Jurassic of Argentina. The Chinese *Bishanopliosaurus* from the Early Jurassic and *Yuzhoupliosaurus* from the Middle Jurassic come from freshwater environments. These fossils show that rhomaleosaurids were not restricted to the Tethys Ocean but also inhabited eastern Panthalassa as well as the Boreal Ocean. Rhomaleosaurids could, it seems, turn up almost anywhere in future, since these finds place them right across the Jurassic world.

The large size and powerful skull of *Rhomaleosaurus* has resulted in it being considered related to pliosaurids (see the discussion on p. 126). Some studies continue to support this view. But pliosaurids share features with animals such as *Plesiosaurus*, including features of the braincase, jaw, pectoral girdle and humerus that are lacking in rhomaleosaurids. Maybe rhomaleosaurids are a distinct early branch on the plesiosaur family tree, only distantly related to pliosaurids.

This giant, wall-mounted replica of the rhomaleosaurid *Rhomaleosaurus cramptoni* is a copy of the original at the National Museum of Ireland in Dublin. It is one of the world's most complete large plesiosaur specimens. The original is lacking only the end part of the left hindlimb.

PLIOSAURIDS

The term plesiosaur is synonymous with the long-necked, small-headed members of this group. But preying on these – and on the other sea-going animals of the Jurassic and Cretaceous – was a group with entirely different proportions, and one that can be considered the most formidable Mesozoic marine reptile group of all: the pliosaurids. They were the ultimate marine predators of the Mesozoic. Reaching 10 m (33 ft) long and perhaps more, equipped with a massive, long-snouted skull and enormous, thick teeth, giant pliosaurids were a constant presence in the world's oceans from the Middle Jurassic until the middle of the Late Cretaceous, a span of around 80 million years. Pliosaurids as a whole, however, originated in the Late Triassic and therefore persisted for around 110 million years.

Pliosaurids have been known since 1841 when Richard Owen named *Pliosaurus brachydeirus* (first spelt *Pleiosaurus*) from the Kimmeridge Clay. Owen was struck by the teeth, which had cutting edges that make them triangular in cross-section. He also noted the shortness of the neck vertebrae, which he compared to those of whales, and the massive limb bones. Other *Pliosaurus* species have been named since and show that members of the genus lived in the Boreal Ocean, the Western Tethys Ocean and deep in the southern hemisphere. A second giant pliosaurid was named in 1873, this time from the Middle Jurassic of France. This is *Liopleurodon*, its name meaning smooth-sided tooth, a reference to the circular cross-section of its teeth. *Liopleurodon* is almost synonymous with the 1999 BBC television series *Walking With Dinosaurs*, in part because the series portrayed *Liopleurodon* as a super-sized 25 m (82 ft) long. Size estimates of this magnitude were proposed by an advisor to the series but are exaggerated, few specimens exceeding 7 m (23 ft).

This photo from 1910 shows the mounted skeleton of the Oxford Clay pliosaurid *Peloneustes*. The hind flippers are obviously larger than the fore flippers. *Peloneustes* is 'abundant' compared to many other plesiosaurs: we know of several skeletons and at least 12 good skulls.

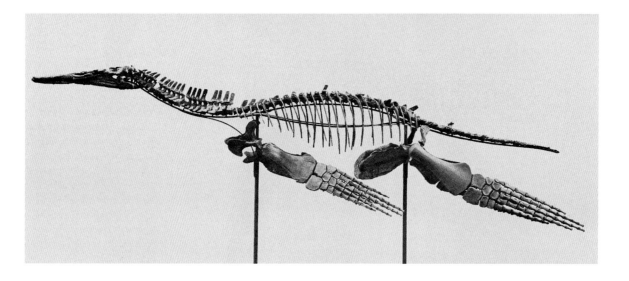

The neck in these animals is short, with just 22 vertebrae, and the eyes are large and show that sight was important. The hind flippers are longer than the fore flippers, the reverse of the normal plesiosaurian configuration. This might have enhanced their ability to accelerate. *Pliosaurus* and *Liopleurodon* were the top predators of their ecosystems. Their teeth are not just robust, pointed and with sharp cutting edges (they belong to Massare's 'cut' guild), but have bulbous roots embedded in deep sockets. The largest teeth are clustered near the tip of the lower jaw and one-third of the way along the upper jaw, these being the regions best at conducting force. In a 1993 analysis of skull function and anatomy, palaeontologists Michael Taylor and Arthur Cruickshank showed that the massive jaw-closing muscles of *Pliosaurus* – enclosed within the enormous, box-like temporal openings – were arranged to allow rapid closing of the jaws and also to clamp them shut and then continue to exert pressure.

These animals surely used their teeth and jaws to subdue prey, which included other plesiosaurs as well as ichthyosaurs, fish and cephalopods. The possibility that they grabbed dinosaurs from coasts or while they were swimming is also plausible. The pliosaurid skull does not, however, possess features that show it was good at withstanding bending and twisting, in contrast to *Rhomaleosaurus*, so they probably avoided wrenching and twisting actions when breaking up prey. Maybe they used cutting and repeated biting instead. Direct evidence for pliosaurid diet comes from stomach contents, which includes fish remains and hooklets from cephalopods, and bite marks. The English Jurassic fossil record has yielded numerous ichthyosaur and

Giant, heavily built pliosaurids like *Liopleurodon* perhaps ambushed prey with a sudden burst of speed. The possibility that they pursued prey over distance also exists. When attacking long-necked plesiosaurs (as shown here), perhaps they aimed to bite the neck or flippers.

Peloneustes is one of several mid-sized, lightly built pliosaurids. It reached 4 m (13 ft) in length. In contrast to giants like *Liopleurodon*, its teeth are slender-crowned and with pointed tips, and its snout and lower jaw are relatively delicate.

This mounted skeleton of *Liopleurodon* – on display at Tübingen University, Germany – was discovered in the English Oxford Clay. Its reconstructed skull is distorted relative to the original. The back of the skull should be taller, and the teeth should be vertically implanted and 'tidier'. Nevertheless, it remains one of the most impressive and influential of large pliosaurid skeletons.

plesiosaur bones marked with scores, scrapes and pits apparently created by the teeth of *Liopleurodon*.

Other pliosaurids lived alongside *Pliosaurus* and *Liopleurodon*. The Oxford Clay fauna includes the shorter-snouted, deeper-skulled *Simolestes*, the slender-snouted *Peloneustes* and also *Pachycostasaurus*, a smaller form around 3 m (10 ft) long, whose short flippers and thick-boned ribs suggest specialization for life close to the sea floor. These co-existing pliosaurids were practising different lifestyles and exploiting different prey.

The fact that animals such as *Peloneustes* were smaller and slimmer-snouted than *Pliosaurus* and *Liopleurodon* reveals a key point. Giant pliosaurids share features of the snout, eye socket, jaw, pelvis and hindlimb lacking in other plesiosaurs except for several plesiosauromorph animals from the latest Triassic and Early and Middle Jurassic. None look like *Pliosaurus* and *Liopleurodon* and some were previously identified as close kin of *Plesiosaurus*. They include the small, Triassic *Rhaeticosaurus* from Germany, the long-snouted *Hauffiosaurus* from Germany and England, and *Attenborosaurus* from Dorset, named after English natural history broadcaster Sir David Attenborough, the original specimen of which was destroyed during WWII (fortunately, casts remain). If the original scientific description of *Attenborosaurus* is accurate, it included a section of skin that revealed a smooth, scaleless surface peppered across the hip region with small, rectangular structures marked with three parallel ridges. Nothing like this has been reported in any other plesiosaur.

All these animals are early pliosaurids. This means that pliosaurids began as small or mid-sized plesiosauromorphs. Only later – mostly after rhomaleosaurids disappeared – did they evolve into the mega-predators that we mostly associate with the name pliosaurid. The large, big-headed pliosaurids are the thalassophoneans, meaning 'sea-going slayers', an apt if slightly dramatic reference to their ecological role.

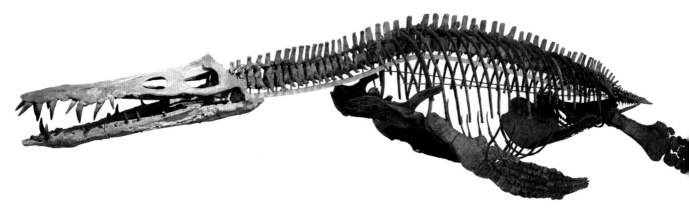

BRACHAUCHENIINES

The story of pliosaurid evolution discussed so far is focussed on Europe. But finds made in the Americas and Australia show that pliosaurids lived in these regions, too. Most of the animals in question belong to a thalassophonean group which differs in several ways from other thalassophoneans. These are the brachaucheniines, named after *Brachauchenius* of Late Cretaceous Kansas and England, its vowel-heavy name meaning 'short neck'. *Brachauchenius* had a skull 90 cm (35 in) long and a total length of between 6 and 9 m (20 and 30 ft). A larger relative – *Megacephalosaurus*, also from Late Cretaceous Kansas – had a skull 1.7 m (5½ ft) long.

The toothed margins of the brachaucheniine upper jaw are straight rather than wavy, the teeth in the upper jaw are mostly the same size and shape, the snout is triangular when seen from above (rather than pinched-in close to the tip), and the jaw joint is especially far back. These features suggest that they could open their jaws particularly wide, which could mean they swallowed large prey. Their style of attack might have been less precise than that of other thalassophoneans. Rather than taking bites from the prey's underside or biting its flippers, perhaps they slammed into it with a wide-open mouth.

The best-known brachaucheniine is from the southern continents. This is *Kronosaurus*, often said to be the largest plesiosaur of all. Part of the reason for its fame is the existence of a reconstructed skeleton at the Harvard Museum of Comparative Zoology in Massachusetts, USA. This was collected (mostly with the aid of dynamite) in Queensland, Australia, during the 1930s and the decision to have it on display came after funding by Boston industrialist Godfrey L Cabot. Cabot did so after quizzing Harvard palaeontologist Alfred Romer about legendary sea monsters, an area in which Cabot had a special interest. Romer is regarded as one of the twentieth century's most accomplished fossil vertebrate experts, but he wasn't a plesiosaur specialist and many of the decisions he made about the reconstruction were wrong. He made the body too long at 12.8 m (42 ft) and gave it too many vertebrae. He also had the skull reconstructed with a tall crest at its rear. Subsequent work has shown that *Kronosaurus* was more like 9 to 11 m (29½ to 36 ft) long, shorter-bodied than the Harvard mount, and with a lower, non-crested skull.

A close relative of *Kronosaurus* – *Monquirasaurus* – was found in Colombia. Its skull is enormous relative to the animal's total length, the lower jaw is almost deeper than the snout, and the body is short and deep. At least three additional giant thalassophoneons, some of which are probably brachaucheniines like *Monquirasaurus*, also inhabited Colombia during the Early Cretaceous. These are the short-snouted *Acostasaurus*, the narrow-nosed *Stenorhynchosaurus* and the especially robust *Sachicasaurus*. Their differences suggest that they were exploiting different prey or hunting in different ways.

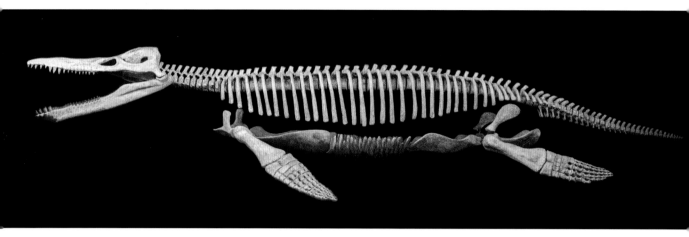

The 'Harvard mount' of *Kronosaurus*. In 2021, it was argued that this specimen (together with other good remains previously labelled '*Kronosaurus*') should be identified as the new species *Eiectus longmani*, and that the name *Kronosaurus* should essentially be abandoned. However, the name *Kronosaurus* is strongly associated with the Harvard mount and changing it now could be seen as disruptive.

All these animals were named between 2016 and 2019 and have only been subjected to preliminary studies so far. Recent finds have also shown that brachaucheniines occurred elsewhere around South America, such as in Venezuela, and that European fossils conventionally labelled *Polyptychodon* can also be attributed to this group, as can *Makhaira* from western Russia, the oldest member of the group reported so far.

It is tempting to conclude from this pattern that brachaucheniines originated in the Western Tethys Ocean before colonizing the Atlantic. This was a new sea at the time, created by a rift between North America and Gondwana. If this is true, brachaucheniine presence is predicted around northern Africa as well as southern and eastern North America. But disrupting this pattern is Australia's *Kronosaurus*, so far-flung that it suggests a global distribution for the group. It implies that brachaucheniines could be found almost anywhere!

A final discovery from the world of brachaucheniines was announced in 2017 – *Luskhan* from the Early Cretaceous of Russia. *Luskhan* was 6.5 m (21¼ ft) long, with a long, narrow snout and small teeth. This shows that brachaucheniine evolution was not just about larger size and capture of large prey, but that at least one lineage was specialized for a diet of small animals and became similar to polycotylids. Plesiosaur evolution was complicated.

PLESIOSAURUS, MICROCLEIDIDS AND KIN

It seems right to regard *Plesiosaurus* – the original plesiosaur – as the most typical member of the group. The problem with this suggestion is that neither it nor any of its close relatives are especially well known. What we can say about *Plesiosaurus* and related forms, grouped together within Plesiosauridae, is that they were not especially big at just 3 to 4.5 m (10 to 15 ft) long, and were generalist predators that captured fish, cephalopods and crustaceans in shallow, continental shelf environments. Their fossils are associated with western Europe. The oldest rhomaleosaurids and pliosaurids come from this region too, so it appears that these groups – and even plesiosaurs as a whole – originated here.

Several lineages belong to the same region of the family tree as plesiosaurids. Among these are the microcleidids, a small, Early Jurassic group also unique to western Europe, named after *Microcleidus*, originally reported from Yorkshire, England. *Microcleidus* has a long, slender neck and a skull where the eyes face partly upwards but also forwards, the teeth are large, thick-based and interlace when the jaws were closed, and the lower jaw is robust and has a reinforced chin. Two tiny teeth are present at the front of the upper jaw. These details suggest a specialized feeding behaviour, but we do not yet have any idea what this involved.

CRYPTOCLIDIDS

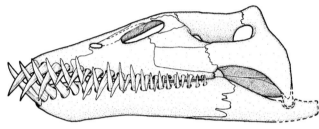
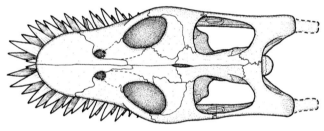

The *Microcleidus* skull reveals an unusual eye orientation and distinctive tooth anatomy. A specimen exists which preserves some of the scleral ossicles that would originally have been embedded within the eyeballs. The slender bone used to conduct sound – the stapes – is also partially preserved in another specimen.

Among the best understood of plesiosaurs are the cryptoclidids, a Middle and Late Jurassic group that persisted into the Early Cretaceous. The group takes its name from *Cryptoclidus*, originally described from the Oxford Clay but also known from France and Cuba. *Cryptoclidus* reached 8 m (26 ft), had a neck of about 32 vertebrae, and possessed around 100 slender, interlocking teeth. Its forelimbs are unusual: the end of the humerus is especially wide and built so that the rest of the limb is angled slightly backwards. The advantage provided by this shape has yet to be identified. The back of the skull is short but tall, with a prominent sagittal crest, while the bones of the cheek are reduced.

Thanks to the discovery of near-complete, three-dimensional skeletons during the late 1800s, mounted *Cryptoclidus* specimens are on show in several museums, including those of London, Glasgow, Paris, Tübingen, Bonn and New York. No other plesiosaur is as well represented in public. Partly as a consequence, *Cryptoclidus* has often served as a model for plesiosaur anatomy, biomechanics and lifestyle. An example comes from Jane Robinson's work of the 1980s, in which it was argued that the vaguely barrel-shaped body of *Cryptoclidus* was key to plesiosaurian swimming style. Robinson suggested that the vertebral column was held perpetually arched and in tension against the gastral basket. How this system is supposed to have worked is not clear, since a vertebral column holds its shape due to the

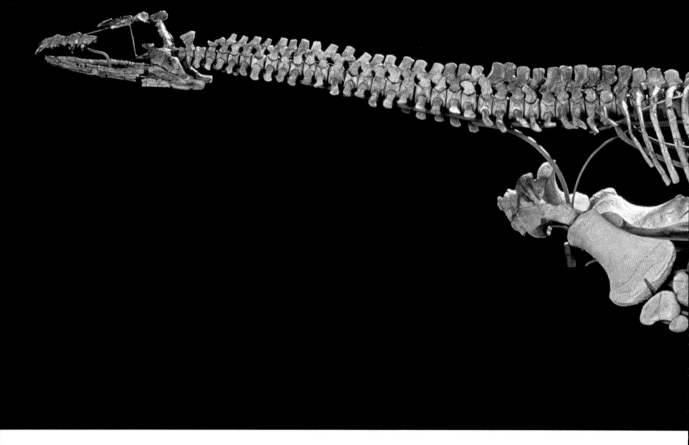

This juvenile *Cryptoclidus* skeleton was discovered in the Oxford Clay of Peterborough, England. As expected for a juvenile, it has a proportionally larger skull than an adult and various of its skull bones had cartilaginous (rather than bony) edges. It has proved important in showing how the bones of these animals, especially those of the flippers, changed during growth.

nature of the joints between the vertebrae, not to tension. Today, it seems that this view on cryptoclidid anatomy is not accurate. In reality, the cryptoclidid spine was straighter and without a humped appearance.

Kimmerosaurus from the Kimmeridge Clay was a small relative of *Cryptoclidus*, and certain details of its skull indicate that it was doing something unusual, though exactly what is yet to be determined. The top of its skull is flattened, rather than in possession of a sagittal crest, and its teeth are recurved and lack the ridges typical for plesiosaurs. Perhaps it is catching smaller, softer prey than *Cryptoclidus*.

The also small (3 m, 10 ft) *Tatenectes* from the Late Jurassic of Wyoming, USA had a broad and flattened body and especially thick, dense-boned gastralia. It's a remote possibility that *Tatenectes* was specialized for lying on the seafloor and that it captured prey by ambush. This is unlikely, however, partly because plesiosaurs were air-breathers, but also because everything about plesiosaur anatomy points to them being active predators. An alternative explanation for the broad, flattened body is that it provided stability during swimming, this perhaps being an adaptation for life in turbulent waters near the coast. Its stomach contents reveal a diet of squid and small sharks.

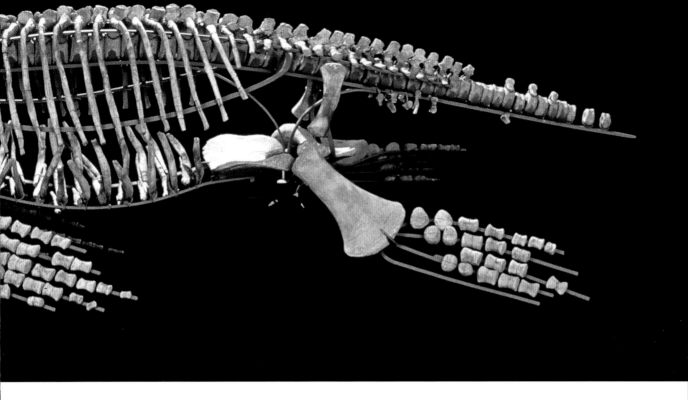

The Jurassic cryptoclidid *Tatenectes* is based on fossils reported in 1900 but not recognized as being especially novel until 2003. It was short-necked relative to other cryptoclidids and with an unusual, flattened body. Some plesiosaurs might have possessed striking markings like those shown here, a real possibility given their excellent vision and probable high degree of sociality.

Abyssosaurus from the Early Cretaceous of Russia, named in 2011, is a poorly known cryptoclidid originally interpreted as an aristonectine. Massive eyes and dense gastralia suggest deep-diving habits. Some experts argue that *Abyssosaurus* was a specialist of cold northern seas.

Recent years have seen the description of additional cryptoclidids, mostly from sediments deposited in a very different environment: namely cool, high-latitude seas. These animals – they include *Spitrasaurus*, *Djupedalia* and *Ophthalmothule* – are mainly associated with Spitsbergen, but some are also known from western Russia. They are mostly 5 to 6 m (16½ to 19¾ ft) long, big-eyed, and with necks that are of elasmosaurid-like proportions (*Spitrasaurus* and *Ophthalmothule* possess more than 50 neck vertebrae). Their jaws are delicate and not suited for the handling of large, muscular or armoured prey, and their large eyes indicate adaptation for life in dark waters. The Russian *Abyssosaurus* also has an especially long, slender neck in addition to a broad body, short tail, atypically long hind flippers and a short-snouted, rounded skull that is especially tall at the back and at the level of the eyes. Its eye sockets are huge. These features indicate that it hunted small fish, crustaceans or squid in deep, dark waters. It might have been one of the most strange-looking of all plesiosaurs. Its unusual proportions and poorly ossified skeleton raise the possibility that its evolution was driven by paedomorphosis (see Chapter 3).

It seems that cryptoclidids were an atypical group, associated with cool, dark, northern seas. What makes this even more interesting is the suggestion from some studies that the species adapted for this environment were not all close relatives, but members of different cryptoclidid clades. There may, therefore, have been a tendency within the group to repeatedly become cold-water deep divers.

ELASMOSAURIDS

Among the most familiar plesiosaurs are the elasmosaurids, a Cretaceous group famous for their exceptionally long necks. A typical long-necked plesiosaur such as *Plesiosaurus* might have 46 neck vertebrae, but elasmosaurid necks typically have 60 vertebrae or even more than 70, making them among the most extreme necks to ever evolve. These animals were also sometimes giant, the biggest reaching 12 m (39 ft). Elasmosaurids are associated with the Late Cretaceous Western Interior Sea but occurred worldwide, and recent studies have shown that their evolutionary history was complicated.

As discussed earlier (see pp. 21–22), *Elasmosaurus* was so unusual that it was initially reconstructed the wrong way round: as a long-tailed animal with a short neck. Interpreted correctly, *Elasmosaurus* was a sensation. Artistic reconstructions of the 1800s show it swimming and hunting, and fighting with mosasaurs. Those produced by pioneering American palaeoartist Charles Knight were especially influential, and shaped views on elasmosaurid appearance, behaviour and biology. Knight showed *Elasmosaurus* throwing its neck into coils and curves, and swimming with it held out of the water in a swan-like pose. Decades of discussion and study has led plesiosaur experts to reject these ideas. Plesiosaur necks were flexible, but even elasmosaurid necks weren't flexible enough to form tight curves and coils, nor could the neck be lifted above the water for reasons of its mass.

In the years following the recognition of *Elasmosaurus* in 1868, other long-necked plesiosaurs were allocated to the same family, including *Microcleidus* from the Early Jurassic of England and *Muraenosaurus* from the Oxford Clay. The classification of these animals as elasmosaurids created the impression that the group persisted mostly unchanged throughout the Mesozoic and that it was conservative, only really varying in neck length and the configuration of the teeth.

Numerous studies published since the 1990s have overturned this. Elasmosaurids were a late-evolving group that underwent most of their diversification after the end of the Jurassic, and it may be that they were unique to the Cretaceous. The idea that they were conservative in anatomy is also not true. They were variable in neck length, their overall proportions, the shape of their skull, teeth and flippers, and probably in feeding behaviour and ecology, too.

Some experts have argued that elasmosaurid proportions demonstrate a lifestyle devoted to the grabbing of small fishes. What seems more likely is that the predatory abilities of these animals have been underappreciated. Their skulls are large, almost 50 cm (19.7 in) long in the biggest species, with huge jaw-closing muscles, strong jaws and massive teeth, the longest of which reached 15 cm (6 in). These features suggest they were formidable predators that tackled prey from 30 cm to 2 m (11¾ in to 6½ ft) long, including cephalopods, large bony fishes, small sharks, diving birds and juvenile mosasaurs. This is supported by stomach contents. These include shark teeth and fin spines, bones from specimens of the fish *Enchodus* 40 cm (15¾ in) long, mosasaur bones and cephalopod parts. Stomach contents from two Australian elasmosaurids include bottom-dwelling molluscs and crustaceans, suggesting that some species were benthic grazers, picking up prey from the seafloor. It is true that much speculation surrounds elasmosaurid feeding behaviour, but they should not be imagined as reliant on sardine-like fish.

Elasmosaurid diversity

The oldest elasmosaurids include species from Russia, England, Germany, Canada and Colombia. The neck vertebrae of these animals are generally shorter than those of later elasmosaurids, and their necks are shorter too, containing less than 60 vertebrae, still more than other plesiosaur groups. Elasmosaurids of this short-necked sort were mostly animals of the first half of the Cretaceous, though a few persisted to its end. Several features beyond neck length are used to characterize elasmosaurids, including the presence of horizontal ridges on the sides of the neck vertebrae (presumably for

The majority of elasmosaurids belong to the Late Cretaceous group Euelasmosaurida. *Albertonectes* (shown here) is the biggest of them all. It exceeded 11 m (37 ft) in total and its neck was 7 m (23 ft) long. Numerous cartilaginous and bony fishes – like the shark, coelacanth and ray-finned species shown here – lived alongside animals like this and fell prey to them.

muscle attachment), a rounded bony projection that enters the lower edge of the eye socket and a heart-shaped space in the middle of the coracoids.

Around 90 million years ago, a new, Late Cretaceous clade evolved, the Euelasmosaurida, meaning 'true elasmosaurids'. This group exploded in diversity and encompassed variation that was new not just for elasmosaurids but plesiosaurs as a whole. Conflicting views exist on how the members of Euelasmosaurida are related to one another. However, the especially long-necked styxosaurines or elasmosaurines certainly belong within this clade. These are associated with the continental seas of North America and include *Elasmosaurus* itself in addition to *Styxosaurus*, also from the USA, and *Terminonatator* and *Albertonectes* from Canada. These were among the largest of all plesiosaurs, and also the longest-necked, their necks sometimes exceeding 5 m (17 ft) in length. *Albertonectes* has the most extreme neck of all, with 75 vertebrae.

It is not the case that all elasmosaurines were super-necked giants, however. *Nakonanectes* from the USA was a small, short-necked elasmosaurine, 5 to 6 m (16½ to 19¾ ft) long and with only 40 or so neck vertebrae. Also atypical is *Cardiocorax* from Angola. Its short flippers, large coracoids and reduced scapulae show that it had an unusual swimming style. Yet again, the fossil record reminds us that plesiosaur evolution was complex, and that few groups were anatomically uniform across their histories.

Styxosaurus

Zarafasaura

Libonectes

Thalassomedon

Elasmosaurids are diverse in skull anatomy, and some members of the group (like *Zarafasaura*) are much shorter-faced than others. Massive jaw muscles and long, interlocking, fang-like teeth are typical. These are not the skulls of animals adapted to eat small, lightly-built prey items.

Aristonectines

Euelasmosaurida also includes the aristonectines, a group that deviated markedly in anatomy and lifestyle from other plesiosaurs. Aristonectines have been known since 1941 when Ángel Cabrera described *Aristonectes* from the Upper Cretaceous of southern Argentina. This was a large plesiosaur with an unusually high tooth count, but little else was clear. Additional aristonectines have been discovered more recently, all dating to the Late Cretaceous. They include *Morturneria* from Antarctica, *Kaiwhekea* and *Alexandronectes* from New Zealand, and *Wunyelfia* from Chile. Additional *Aristonectes* specimens have been discovered in Chile and Antarctica. This was a group with strong ties to the cool waters of the far south.

Aristonectines were among the most specialized of plesiosaurs, their distinctive traits including multi-toothed jaws, massive flippers and a robust tail with a horizontal tail fluke. This reconstruction depicts the Antarctic aristonectine *Morturneria*.

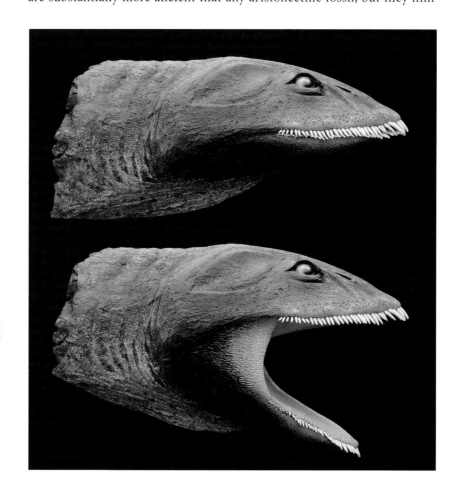

Cabrera noted a similarity between the jaws of *Aristonectes* and those of baleen whales. The more complete specimens known today show that aristonectines had a flattened skull where the back of the jaws was wide and able to engulf a large quantity of water. The slender teeth project outwards from the jaws and form a comb-like arrangement that could function as a filter when the jaws were held slightly open. It is tempting to suggest that aristonectines were feeding on plankton. However, the spacing between their teeth is not fine enough for this, so an alternative suggestion is that they were sediment-sievers, who took mouthfuls of mud and then filtered the animals out of it. Small crustaceans, molluscs and other invertebrates lived in and on the sediment in the environments where aristonectines occurred. A section of petrified Middle Jurassic seafloor from Switzerland preserves a number of grooves, gutters and cavities that appear to have been made by plesiosaurs and other marine reptiles as they foraged in the sediment. These feeding traces are substantially more ancient that any aristonectine fossil, but they hint

A live aristonectine would have had a highly unusual body plan when viewed from above or below like this, the relatively stout neck, exceptional flipper span and tail fluke being among their key traits. The reconstruction shows the bones known from this one particular specimen, but other aristonectine fossils are even more complete.

These reconstructions of the aristonectine head (specifically showing *Morturneria*) show how the teeth did not interlace (as is typical for plesiosaur skulls) when the jaws were closed. Instead, the upper and lower jaw teeth would have projected in parallel.

at the idea that plesiosaurs of several groups fed on mud-dwelling animals throughout the Mesozoic.

It is not just the aristonectine skull that is odd, the rest of their anatomy is, too. The neck is thick and deep. The limbs are exceptionally long, in cases about a third of the body length, and the shoulder and hip joints indicate substantial limb flexibility. Flattened vertebrae show that the tail fluke was horizontal. Finally, some aristonectines were enormous, and among the biggest plesiosaurs of all. *Aristonectes* reached 10 m (32¾ ft) and its long flippers gave it a span of 7 m (23 ft). One member of the group from Antarctica, not yet known from remains good enough to warrant a name, reached 12 m (39½ ft) long.

Aristonectines have similarities with early baleen whales. It is impossible to say what would have happened to aristonectines had they continued evolving beyond the Cretaceous, but it is tempting to suggest that their unusual proportions, large size and adaptations for sieve-feeding could have resulted in plesiosaurs of even more remarkable form.

LEPTOCLEIDIDS

Southern England has yielded numerous fossil marine reptiles, and the majority come from sediments deposited in marine environments. But in 1922, English palaeontologist Charles W Andrews described a new kind of plesiosaur. It came from the Lower Cretaceous Wealden of Sussex, a group of mudstones, sandstones and siltstones deposited in freshwater and estuarine environments. Rather than being a denizen of coasts or the open sea, Andrews' plesiosaur inhabited estuaries, river mouths and lagoons. This is *Leptocleidus*, its name meaning 'slender clavicle', a reference to its narrow, bar-like collarbones.

Andrews was struck by the environmental setting of *Leptocleidus* and combined it with an interpretation of the animal as primitive relative to other Cretaceous forms. Perhaps, he suggested, it was a 'living fossil', a relict that led "a life sheltered from the great competition that seems to have resulted

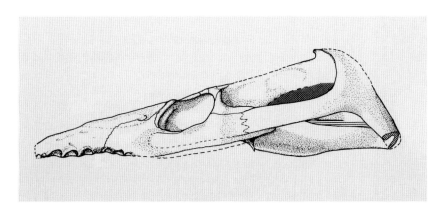

The leptocleidid skull is 'old fashioned' in shape but possesses unusual details. The length of the section behind the eyes is unusual, as is the crest that extends along the mid-line of the upper surface. This is the skull of the South African species *Leptocleidus capensis*.

in more rapid evolution among the marine plesiosaurs". He made direct comparisons with river dolphins such as those of the Ganges River in India. River dolphins descend from marine ancestors, and the idea has long been popular that they avoided competition with evolutionarily younger dolphins by persisting in rivers and lakes. This view is old-fashioned today, but was thought correct in the 1920s. *Leptocleidus* is a mid-sized, pliosauromorph plesiosaur, 3 m (10 ft) long, with 20 to 25 neck vertebrae. The snout has a squared-off tip and the part of the skull behind the eyes is long. A low bony crest is present along the skull's midline. When alive, *Leptocleidus* might have looked like a miniature rhomaleosaurid (hold that thought!).

The Wealden *Leptocleidus* species is not unique. Even when Andrews was describing the specimen in 1922, similar animals were known from the estuarine and lagoonal sediments of South Africa, and later discoveries were made in Western Australia. *Leptocleidus* was thus widely distributed. However, the estuarine and freshwater environments of these regions were never directly connected, so ancestral *Leptocleidus* populations must have made sea crossings during their history. Presumably they moved along the northern and eastern coastline of Africa, which at the time was connected to the other Gondwanan continents, before giving rise to the English, South African and Australian species.

Several similar animals also inhabited freshwater and estuarine environments during the Cretaceous. All have intermediate neck proportions, short neural spines and an unusual configuration of bones on the skull roof: specifically, the frontal bones have extra sections that reach back to make contact with the temporal openings. All are united within the group Leptocleididae. *Brancasaurus* from the Early Cretaceous of Germany – the best understood European Cretaceous plesiosaur – might be a leptocleidid, though an elasmosaurid identity has also been suggested. *Brancasaurus* has a longer neck than *Leptocleidus*, with 37 vertebrae, and distinctive flattened bones between the eyes, the upper surfaces of which possess an odd dagger-like structure formed of raised bony ridges. The many specimens of *Brancasaurus* come from estuarine and lake environments, specifically from a giant lake that was fresh for part of its history but inundated by seawater at other times. Many of these individuals are juveniles, so it has been suggested that the area was a plesiosaur 'nursery'. It cannot have been an entirely safe nursery, however, since a *Simolestes*-like pliosaurid is known from the same location.

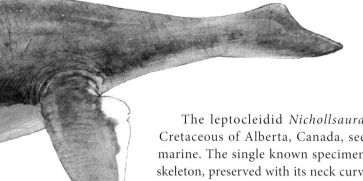

The leptocleidid *Nichollsaura* from the Early Cretaceous of Alberta, Canada, seems to have been marine. The single known specimen is an articulated skeleton, preserved with its neck curved to the side in a life-like pose. *Umoonasaurus* from the Early Cretaceous of southern Australia also appears to be a marine leptocleidid, and one that inhabited cold seas where winter temperatures were below freezing. *Umoonasaurus* has the crest typical of the group, but bony arches over its eyes have been interpreted as additional crests, perhaps horn-covered in life. This is likely a mistake, and these are probably the bony rims over the eye sockets seen elsewhere in plesiosaurs.

The lightly built palate and small, slender, gently curved teeth of leptocleidids suggest reliance on small prey. Their small size, mid-length necks and presence in estuaries and freshwater suggest that they were good at foraging in narrow channels and along cluttered banks. The marine forms might have been animals of coastal inlets and fjords. They failed to persist beyond the middle of the Cretaceous, around 100 million years ago, although the reason for their disappearance is unknown.

Because leptocleidids have pliosauromorph proportions, the view for much of the late twentieth century was that they are pliosaurids. Leptocleidids recall rhomaleosaurids in several details, a result being the suggestion that they were dwarfed, late-surviving descendants of that group. The bulk of evidence, however, shows that they belong within Xenopsaria.

> Leptocleidids appear to have been mid-sized generalists adapted for life in 'marginal' habitats like lagoons, estuaries and lakes, not the open ocean. Such environments encourage behavioural flexibility, small size, and an ability to be a manoeuvrable swimmer. This reconstruction depicts the geographically widespread *Leptocleidus*.

POLYCOTYLIDS

Our tour of plesiosaur groups has shown how the short necks and large, long-snouted heads associated with the pliosauromorph shape evolved three times: in rhomaleosaurids, pliosaurids, and again in leptocleidids. A fourth pliosauromorph group exists and it is one of the most modified, unusual plesiosaur groups of all.

These are the potycotylids named in 1909 by American palaeontologist Samuel Williston after *Polycotylus*, an animal first reported from the USA but today known from Russia too. *Polycotylus* was thus probably widespread in Late Cretaceous seas. It was mid-sized at 5 m (16 ft) long, with a long, narrow, shallow snout and conical teeth. The rear of its palate is wide and extends surprisingly far back. The flippers are long, and extra bones and interlocking joints show that they were stiffer than those of other plesiosaurs.

Many polycotylids have been named since *Polycotylus*'s debut. Most are from the Late Cretaceous of the USA, but others are from Canada, Australia,

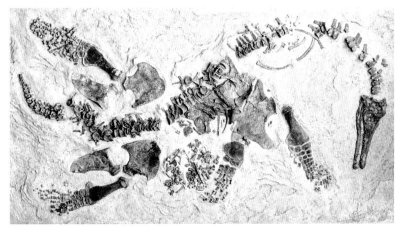
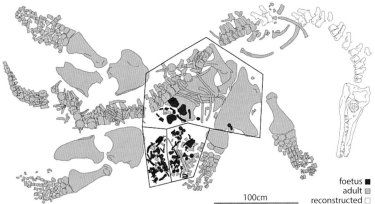

foetus ■
adult ▨
100cm   reconstructed ☐

This *Polycotylus* specimen – on show at the Natural History Museum of Los Angeles, USA – contains the remains of a foetus inside its skeleton. For now, we assume that the production of proportionally enormous babies, born live at sea, was common to all plesiosaurs.

New Zealand, Japan, eastern Europe, Morocco, Argentina and Mexico. Features that unite them include a lower jaw where bones normally restricted to the rear extend to the front, a nostril that no longer has a contact with the premaxillary bone, and short neck vertebrae.

The oldest polycotylids – such as *Edgarosaurus* from Montana – are from the late Early Cretaceous, and are around 100 million years old. The close relationship between polycotylids and leptocleidids, however, means that polycotylid history must extend back to the start of the Cretaceous, since we know of leptocleidids at least that old. *Hastanectes* from the Wealden in southern England might be an early polycotylid and is from an early point in the Cretaceous, but is unfortunately only known from vertebrae.

By the start of the Late Cretaceous, several polycotylid groups had evolved. A subgroup known as the occultonectians evolved from an *Edgarosaurus*-like ancestor. Members of this group are known from the USA, Argentina and Australia. The Argentinian occultonectian *Sulcusuchus* has grooves on its palate and jaws that perhaps housed sensory organs. Similar structures are seen elsewhere in plesiosaurs, so perhaps features of this sort were typical.

The name 'occultonectian' means 'hidden swimmer' and refers to the fact that the members of this group were misidentified as non-polycotylids in the past, sometimes even as non-plesiosaurs. The general trend in polycotylid evolution was to increase the length of the snout, but occultonectians reversed this. They also evolved prominent ridges on their teeth – a feature associated with the grasping of large prey – and appear to have been powerful, pliosaurid-like predators. One occultonectian – *Plesiopleurodon* from the Late Cretaceous of Wyoming – was initially misidentified as a *Liopleurodon*-like pliosaurid.

Another group – the polycotylines – originated at about the same time as the occultonectians. Lineages within this group went in different evolutionary directions. Some became long-necked: in *Thililua* from Morocco, the neck is more than three times longer than the crocodile-like skull. Others became short-necked: in *Mauriciosaurus* from Mexico the neck is shorter than the skull. Yet others became slender-snouted and small: *Dolicorhynchops* and *Trinacromerum* were only 3 m (10 ft) long. This diversity again gives some idea of how adaptable and anatomically flexible plesiosaurs were. It wasn't difficult for them to modify size, snout proportions and neck length. This shows that we shouldn't rely on those features when allocating specimens to specific groups.

A 2018 study devoted to polycotylid history, led by marine reptile expert Valentin Fischer, found that polycotylids underwent their greatest evolutionary diversification around 95 million years ago. This is when the Cenomanian age of the Late Cretaceous ended and the Turonian age began, a time when marine ecosystems were being reorganized due to changing global conditions. Both occultonectians and polycotylines persisted to the end of Cretaceous times.

Diverse snout and tooth shapes, neck lengths and body sizes show that polycotylids occupied diverse niches. The long, stiffened flippers and robust, broad-based snouts of forms such as *Polycotylus* suggest that they were fast-moving, agile predators equipped to grab mid-sized prey. Others – such as the occultonectians – were top predators of large animals that probably relied on ambush and powerful, crushing bites. In contrast, the small size, slender jaws and long, slim teeth of polycotylines such as *Trinacromerum* suggests that they were predators of small, fast-swimming fishes and cephalopods. Stomach contents from a Japanese polycotyline reveals that ammonites were sometimes eaten. A few polycotylid specimens are preserved with intact gastrolith masses, in some cases around 300 of them. The long-necked, slender-snouted polycotylines such as *Thililua* remain enigmatic and we have no clear idea of how they lived. One possibility is that they used the neck and long, slender jaws to probe into crevices.

One *Polycotylus* specimen from the Late Cretaceous of the USA is among the most important plesiosaur fossils of all in terms of what it tells us about their biology. The specimen concerned is a mother, and she was pregnant when she died. The foetus she contains is enormous and would have been around 40 per cent of its mother's length when born. Its bone texture shows that it was growing quickly, a discovery in keeping with evidence that plesiosaurs had a high metabolism.

Reptiles that produce exceptionally large young tend to invest time and effort into them once they are born. After all, it makes evolutionary sense to assist in the survival of an offspring that has already 'cost' the mother a great deal. It therefore seems likely that *Polycotylus*, and presumably other polycotylids and xenopsarians, and perhaps all plesiosaurs, practised parental care, the mother and her offspring living together for an extended time, perhaps in a family group.

7 | SEA CROCS:
THE THALATTOSUCHIANS

This reconstructed skeleton of *Pelagosaurus* from the Early Jurassic of western Europe betrays features typical of many thalattosuchians. These include slender jaws, small limbs and lightweight proportions. *Pelagosaurus* is mid-sized, at 2–3 m (6½–10 ft long). Later thalattosuchians evolved much greater size and substantially more specialized anatomies.

LIVING ALONGSIDE ICHTHYOSAURS and plesiosaurs in the seas of the Jurassic and Cretaceous were the thalattosuchians, sometimes called sea crocodiles or sea crocs. Thalattosuchians were not crocodiles at all but part of the much larger group – known technically as Crocodylomorpha – that includes today's crocodiles, alligators and gharials as well as a large number of extinct crocodile-like reptiles.

Crocodylomorph diversity is substantial, and the group's history is complex and confusing. Crocodylomorphs originated around 230 million years ago during the Triassic, and among their oldest members are a variety of lightly built, terrestrial predators. These possessed lightly built tails and long, erect limbs and bore little resemblance to modern crocodiles. Animals of this sort gave rise to numerous additional crocodylomorph groups throughout the Jurassic, Cretaceous and much of the Cenozoic, including omnivores, predators and herbivores, large and small. During the Jurassic, a new group of crocodylomorphs evolved, characterized by amphibious habits, 'wavy' edges to the upper jaw, and two clusters of enlarged teeth. These are the neosuchians, the group that ultimately gave rise to modern crocodiles, alligators and gharials (all of which are grouped together as the crocodylians).

This exceptional *Rhacheosaurus* skeleton from the Jurassic of Germany highlights the narrow, lightly built form of certain metriorhynchids. This specimen also preserves a soft tissue outline, including a tail fluke.

The skull of the Jurassic metriorhynchid *Gracilineustes*. Shallow, lightly built jaws and slim, fine teeth were present in many thalattosuchians. These animals were predators of squid and small fishes. However, other members of the group became adapted for very different prey.

Several crocodylian species – most famously the saltwater crocodile *Crocodylus porosus* of southern Asia and Australasia – have become sea-going animals, but none have become as fully adapted to marine life as the Mesozoic thalattosuchians. Two main thalattosuchian groups exist. The first – the teleosauroids – are superficially similar to crocodylians and were recognized as close kin of gharials or crocodiles when discovered in the early 1700s. The second – the metriorhynchids – are more modified. Metriorhynchids lack bony armour on the surface of the body, possess paddles in place of clawed limbs, and have a downcurved tail-tip that supported the lower lobe of a tail fin.

THALATTOSUCHIANS IN THE CROC FAMILY TREE

Where thalattosuchians belong within Crocodylomorpha is controversial and several competing possibilities have been suggested. The view favoured for most of the late twentieth century was that thalattosuchians are neosuchians, and closely related to several long-jawed neosuchian groups of the Cretaceous and Cenozoic.

Recent studies have tended to place thalattosuchians some distance away from neosuchians, and in fact outside the clade that contains all crocodylomorphs similar to neosuchians, or close to neosuchian ancestry. This would mean that thalattosuchians evolved early in crocodylomorph

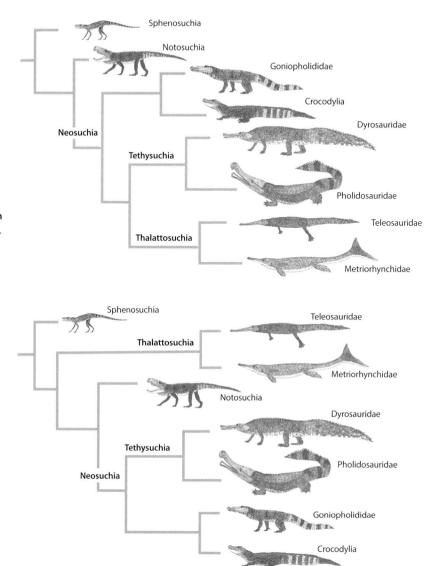

Studies disagree on where thalattosuchians belong in the crocodylomorph family tree. Two possibilities are shown here. In one tree, thalattosuchians evolved late in crocodylomorph history and are close to other long-jawed groups. In the other tree, thalattosuchians evolved early in crocodylomorph history and are not close kin of other long-jawed groups.

Teleosauroids (opposite) are often compared to modern crocodylians, in particular gharials. However, their anatomy is substantially more specialized. Key features include a proportionally large head and unusual limb proportions relative to other crocodylomorphs, and a unique internal skull structure.

history from long-legged, terrestrial crocodylomorphs, not from amphibious or aquatic crocodylomorphs of the sort ancestral to neosuchians. It would also mean that thalattosuchians are only similar to neosuchians due to convergent evolution.

This scenario also means that we have yet to discover, or at least recognize, fossil animals that bridge the evolutionary gap between ancestral, terrestrial crocodylomorphs and the oldest thalattosuchians. Unfortunately, it appears that thalattosuchians originated during a part of crocodylomorph history that is not well understood, and where the fossil species we know of are mostly specialized members of the respective groups.

TELEOSAUROIDS

Of the two main groups within thalattosuchians, we begin with the teleosauroids of the Jurassic and Early Cretaceous. Teleosauroids have traditionally been regarded as a conservative group of superficially gharial-like fish-eaters adapted for coastal waters, some of which were big enough and with sufficiently robust teeth to prey on shelled animals such as turtles. However, recent studies have demonstrated great diversity among teleosauroids, a complex evolutionary history, and the requirement for a flurry of new taxonomic names.

Teleosauroids take their name from *Teleosaurus*, a Middle Jurassic crocodylomorph originally described from France. This was first named in 1820 but treated initially as an extinct species of modern crocodile, and it took a later study of 1825 to correct this mistake and give it the name we use today. *Teleosaurus* is one of the earliest fossil marine reptiles known to science. A long, narrow rostrum lined with small, slender teeth give it proportions similar to those of gharials. Later finds revealed the presence of this group of animals in the Jurassic rocks of England (Dorset and Yorkshire), continental Europe (Germany, Luxembourg, Belgium, Portugal, Switzerland, Slovakia and Poland), Russia, Africa (Morocco, Ethiopia and Madagascar) and Asia (China, Thailand and India). These fossils are from sediments deposited in coastal environments, but Jurassic teleosauroids from eastern Asia inhabited rivers and lakes. These freshwater teleosauroids presumably evolved from marine ancestors that moved into estuaries and rivers. Similar evolutionary transitions happened numerous times in fish, plesiosaurs and marine mammals.

Several teleosauroids are known from complete skeletons, preserved with their osteoderms in place and with patches of scaly skin preserved. Their limb proportions and body shapes suggest an ability to walk on land, and the presence of large osteoderms is consistent with life on and around coasts – the osteoderms provided protection from predators such as predatory dinosaurs, perhaps provided attachment sites for muscles used in walking, and were used to collect heat when the animals basked on land. Teleosauroids also possessed similar jaw muscles that crocodylians use to hold their mouths open while basking, a behaviour linked to temperature control.

Aeolodontines and machimosaurids

Teleosaurus and other 'typical' teleosauroids belong within a group called teleosaurids. A subgroup within the teleosaurids are the aeolodontines of Late Jurassic England, France and Germany. These possess smaller, thinner, more lightweight osteoderms than other teleosaurids, and more streamlined skulls. Aeolodontines are among the most longirostrine of all crocodylomorphs, and most have very small forelimbs. The fossils of one species come from sediments deposited in deep water. Combined, these features show that aeolodontines were semi-pelagic, and very different ecologically from the coastal teleosaurids.

A number of teleosauroids are more heavily built and larger than those discussed so far. These are the machimosaurids, named after *Machimosaurus* of the Late Jurassic and Early Cretaceous of western Europe and Tunisia. *Machimosaurus* teeth are blunt-tipped and rounded (they fall into Massare's 'crunch' guild, see pp. 40–41), the rostrum is broad, almost as wide as the skull is across its eye sockets, and the jaws are heavily built. *Machimosaurus* was gigantic, and its biggest species– such as the Tunisian *M. rex*, named in 2016 – exceeded 7 m (23 ft). This array of features shows that machimosaurids were powerful predators of large animals, their prey including turtles, other marine reptiles, and possibly dinosaurs that came to the water's edge or were grabbed while swimming. Direct evidence for machimosaurid diet comes from turtle shell fragments and dinosaur bones that possess *Machimosaurus* bite marks and broken tooth fragments on their surfaces.

Prior to 2016, machimosaurids were assumed to be a small, evolutionarily conservative group. However, a series of post-2016 studies led by teleosauroid specialist Michela Johnson have shown that several Middle and Late Jurassic fossils – some previously misidentified as specimens of *Teleosaurus* – are additional machimosaurids, distinct from the known kinds and requiring new names. They differ in the characteristics of their snouts, jaws and limbs, and appear to have been specialized for different prey and different roles in their ecosystems. They include the Madagascan *Andrianavoay*, the exceptionally long-snouted *Charitomenosuchus*, and the massive *Lemmysuchus* from the Oxford Clay.

As is typical for Mesozoic marine reptile groups, our knowledge of machimosaurids – and indeed teleosauroids in general – is dominated by the fossil record of western Europe. The fringes of Africa and southern Asia have a lesser but critical role. During the Jurassic and Early Cretaceous, these animals were clearly denizens of the Western Tethys Ocean, even present far to the south, as evidenced by their presence in Ethiopia, Madagascar and India. However, there appears no good reason why they couldn't also have occurred further west, around the edges of North and South America in the newly opening North Atlantic. It might be that their absence here is an artifact of the patchiness of the fossil record.

> The gigantic teleosauroid *Machimosaurus rex* is one of the largest teleosauroids, and also one of the youngest in geological terms since it was alive during the Early Cretaceous. This animal could well have been an ambush predator that captured large, land-living animals like dinosaurs.

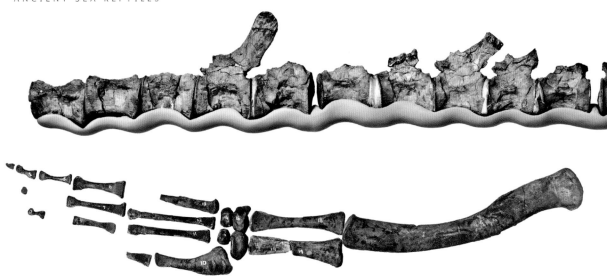

Metriorhynchid hind limbs are unusually proportioned: the foot is long and built to form a flipper, the shin is short, and the thigh is long and built to allow considerable flexibility at the hip joint. The tail has a modified end and supported a vertical fluke. These fossils belong to the European Jurassic metriorhynchid *Gracilineustes*.

Evidence that this is so comes from possible Middle Jurassic teleosauroid fossils from Oregon, USA, as well as possible material from the Late Jurassic of Mexico and a partial machimosaurid skeleton from the Lower Cretaceous of Colombia. The Colombian remains consist of vertebrae, osteoderms and limb bones from an enormous animal, estimated to have been 9.6 m (31 ft) long. This is the biggest teleosauroid and thalattosuchian known, and hints at the possibility that teleosauroids of the far west could prove unusual and surprising relative to their cousins from further east. Future finds will show whether teleosauroids were present more widely on the southern shores of the North Atlantic, and perhaps on its northern shores, too.

METRIORHYNCHIDS AND THEIR KIN

Teleosauroids were highly modified for the marine environment, but even they were relatively conservative compared to their entirely marine cousins, the metriorhynchids. Metriorhynchids shared an ancestor with teleosauroids, and both diverged during the Early Jurassic. The metriorhynchid skeleton was suited for life at sea, their bones being lightened due to the evolution of a spongy interior. Their limb girdles are reduced, and their limbs are specialized. The main bones of the forelimb are semi-circular and plate-like, meaning that the arm was flattened and paddle-like. Metriorhynchid forelimbs are small in some species and might only have had a role in keeping the animal stable while swimming, though a role in steering is also possible.

Metriorhynchid hindlimbs combine a long femur with a shortened lower leg and elongate, paddle-like foot. It appears that the hindlimbs were highly flexible and that the feet were used to generate thrust when an extra burst of power was needed. The metriorhynchid tail is remarkable relative to that of other crocodylomorphs due to the specialized anatomy of its end

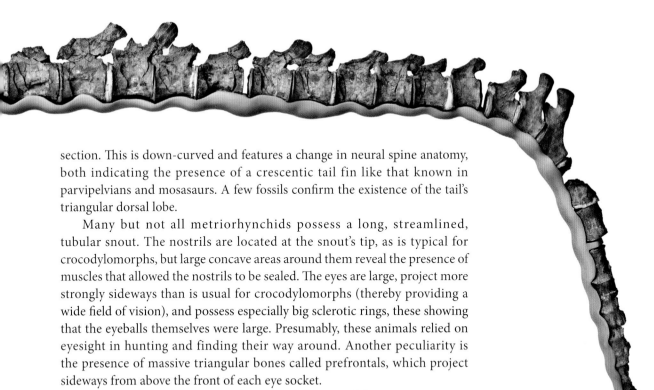

section. This is down-curved and features a change in neural spine anatomy, both indicating the presence of a crescentic tail fin like that known in parvipelvians and mosasaurs. A few fossils confirm the existence of the tail's triangular dorsal lobe.

Many but not all metriorhynchids possess a long, streamlined, tubular snout. The nostrils are located at the snout's tip, as is typical for crocodylomorphs, but large concave areas around them reveal the presence of muscles that allowed the nostrils to be sealed. The eyes are large, project more strongly sideways than is usual for crocodylomorphs (thereby providing a wide field of vision), and possess especially big sclerotic rings, these showing that the eyeballs themselves were large. Presumably, these animals relied on eyesight in hunting and finding their way around. Another peculiarity is the presence of massive triangular bones called prefrontals, which project sideways from above the front of each eye socket.

What might be the most profound modification of this group concerns the lack of osteoderms. Ordinarily in crocodylomorphs, osteoderms provide protection from predators and competitors and also appear to function as collectors of solar heat. It has also been argued that they serve an important function in terrestrial locomotion, since muscles attached to their undersides and to the spine seemingly form a gravity-fighting system that helps the animal hold its weight off the ground. This is called the 'tragsystem' or self-carrying system. It is assumed that osteoderms were lost as a streamlining and weight-saving adaptation, presumably in those thalattosuchians ancestral to metriorhynchids. Several such animals are known, including *Opisuchus* from the Middle Jurassic of Germany and *Zoneait* from the Middle Jurassic of Oregon. These are more similar to metriorhynchids than to other thalattosuchians but lack key metriorhynchid features, like paddle-shaped limbs and sideways-facing rather than slightly upwards-facing eyes. These 'near-metriorhynchids' feature a mosaic of anatomical features. *Magyarosuchus* from the Early Jurassic of Hungary, for example, possessed osteoderms on its upper and lower surfaces, but has tail vertebrae suggesting the presence of a small tail fin.

The fact that metriorhynchids proper lacked osteoderms has become associated with the idea – mostly conveyed through artwork – that they lacked scales, and perhaps had smooth, dolphin-like skin. As revealed by Frederik Spindler and colleagues in 2021, fossil skin segments from metriorhynchids do indeed reveal a smooth skin, albeit with creases, folds and small scars and circular irregularities probably caused by parasites.

SALT GLANDS, TEMPERATURE CONTROL, AND REPRODUCTION

In 2000, Argentinian palaeontologists Marta Fernández and Zulma Gasparini reported the discovery of unusual paired structures within the skull of an Argentinean metriorhynchid. These structures have a lumpy surface texture and resemble the salt-excreting glands present in the heads of seabirds. This is almost certainly what they are. Marine reptiles of all sorts had or have specialized glands to remove excess salt, and metriorhynchids are one of the few fossil groups where they're preserved. In metriorhynchids, these glands were located within the snout interior. A duct connected each gland to a bony preorbital opening on the sides of the snout. Once unwanted salt is excreted, there needs to be a way for it to be transported away from the body. Perhaps the salt secretion, once located within the preorbital opening, was removed via the action of swimming, water movement across the animal's snout sucking the solution outwards. Alternatively, perhaps soft tissue structures in the preorbital opening forced the solution outwards, possibly via an explosive squirt.

One skull detail worth mentioning is that these animals lacked the opening normally present at the rear of the crocodylomorph lower jaw, termed the mandibular fenestra. Ordinarily, this opening provides space for a muscle that crocodylians and, presumably, extinct crocodylomorphs too, used to hold their jaws open – a behaviour termed gaping – when basking on land. The muscle and mandibular fenestra therefore play a part in temperature control. The fact that metriorhynchids lacked this opening shows that they did not gape, and therefore did not bask. How, then, did they control their body temperature?

In recent years several studies have used oxygen isotopes preserved in the bones and teeth of fossil reptiles to determine the temperatures of the animals in life. Metriorhynchid data shows that their temperatures were higher than those expected for crocodylomorphs, and that they could generate at least some heat internally: they were partially 'warm-blooded', though not as proficient at generating heat as parvipelvians and plesiosaurs. Presumably, this helps explain how metriorhynchids survived in Jurassic and Early Cretaceous seas that were cool or even cold (see pp. 14–16). Furthermore, the fact that metriorhynchids appear to have been absent from truly cold waters – like those of the Boreal Ocean – is consistent with them not being as cold-tolerant as parvipelvians and plesiosaurs.

These animals look so similar to plesiosaurs and ichthyosaurs in their degree of marine specialization that we have to wonder whether they were unable to move on land. If crocodylomorph osteoderms really do play an important role in muscular support and movement on land (see p. 159), their absence in metriorhynchids could be linked to the fact that these animals never ventured onto land at all.

We know for sure that marine reptiles belonging to all the main groups (except sea turtles) were viviparous. Might the same have been true of metriorhynchids? A few clues suggest that it was. The metriorhynchid pelvis is odd, being deepened and with a taller rear opening than is typical for crocodylomorphs. This suggests a habit of giving birth rather than laying eggs. Are pregnant specimens known? As yet they are not. However, a tiny juvenile *Dakosaurus* specimen from Germany possess anatomical features highly suggestive of live birth. Remarkably, it lacks ossified forelimbs and cannot have moved across land. It must, therefore, have been born in the water.

> Experts had long predicted the presence of salt-excreting glands in Mesozoic marine reptiles but direct evidence for their presence is rare. Several metriorhynchid fossils have the glands preserved in place within the skull's interior.

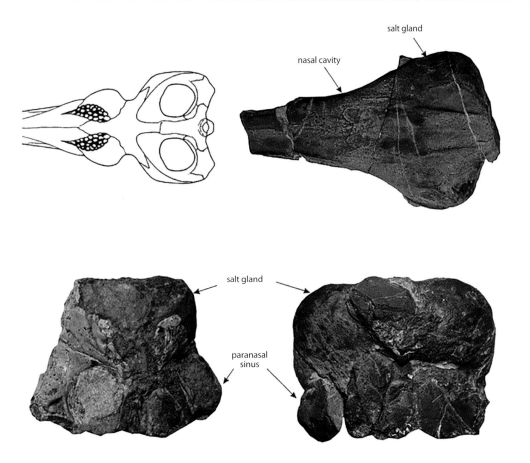

Metriorhynchid diversity

Metriorhynchids have been known since the late 1700s, the oldest specimens being – as is typical for fossil marine reptiles – from the Middle and Late Jurassic rocks of western Europe, France in particular being important. By the 1830s both the deep-snouted *Geosaurus* and the shallower-snouted *Metriorhynchus* were known. Numerous additional specimens were described later, first from Italy, England and Germany but later from Switzerland, Poland, Russia, Argentina, Chile, Mexico, Cuba, Spain, Czechia and Slovakia. The scientific history of the group is complicated, but an enormous amount of recent work – much of it led by Scottish thalattosuchian expert Mark Young – has overhauled our view of this group. It is now clear that the seas of the Jurassic and Early Cretaceous were occupied by metriorhynchids belonging to more than 12 genera and two major subgroups.

The first of these is the narrow-snouted Metriorhynchinae, named after *Metriorhynchus*. Most species ranged from 2 to 4 m (6½ to 13 ft), and most were lightly built, shallow-bodied and slender, with a skull where the snout and jaws are shallow and narrow relative to the back of the skull. Their teeth are slim, pointed and adapted for the grabbing of small squid and fishes, and they were probably swift, agile predators of fast-moving prey. Numerous species have been named.

The second main metriorhynchid group includes larger, more robust animals with deeper, broader, more powerful skulls and larger teeth. These are the geosaurines, a group that originated in the Middle Jurassic and persisted throughout the Early Cretaceous. Geosaurines were especially diverse in western Europe, the majority known from Germany, France and the UK. They are also known from Chile, Mexico and Argentina. It is not yet possible to work out whether this group originated in the Western Tethys Ocean or the Pacific since equally old species are known from both regions. It seems odd that they didn't also colonize the Pacific coast of North America or move east to colonize the seas south of China. Maybe they just didn't, or maybe they did and we have yet to find fossils to prove it.

The geosaurine skull is deep and compressed from side and side. Massive, serrated, recurved teeth line the jaws. In some geosaurines, flattened, worn surfaces were produced as the upper and lower teeth slid against each other during jaw movement, the result being sharp edges effective at slicing into prey. These animals were big, the smallest (like some species of *Geosaurus*) reached 3 m (10 ft), the largest (like *Plesiosuchus*) exceeded 6.5 m

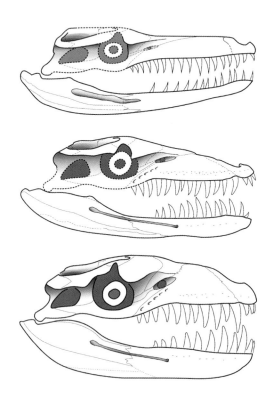

Metriorhynchids are diverse in skull anatomy. Geosaurines – shown here – have heavily built, relatively short snouts and jaws. From top to bottom, these images show *Plesiosuchus* from western Europe and the *Dakosaurus* species *D. maximus* from western Europe and *D. andiniensis* from Argentina.

Geosaurus, 3 m (9¾ ft)

Dakosaurus, 4.5 m (14¾ ft)

Torvoneustes, 4.7 m (15½ ft)

Plesiosuchus, 6.8 m (22¼ ft)

7 m (23 ft)

(21 ft). Geosaurines were powerful predators of large vertebrates, their prey including ichthyosaurs, plesiosaurs and other marine reptiles, in addition to large fish.

Until recently, it was thought that the Cretaceous was a time of gradual metriorhynchid decline, and that they went extinct around 132 million years ago during the Early Cretaceous. This decline was linked to the global changes of this time (see Chapter 2), or perhaps the evolution of new fish groups. However, twenty-first century discoveries show that Cretaceous metriorhynchids were more diverse than previously thought and were still present around 120 million years ago. In fact, they were still diversifying: species with especially big eyes and big tail fins indicate that the Cretaceous members of some lineages were doing entirely new things, such as foraging in the mesopelagic or twilight zone. For all this success, the group was extinct before the end of the Early Cretaceous.

> Metriorhynchids were mostly between 2 and 4 m (6 and 12 ft) long but giant members of the group – like *Plesiosuchus* – reached nearly 7 m (23 ft). These large predators often lived alongside one another in the same environments. Variation in tooth and jaw anatomy, and in size, show that they were exploiting different prey.

8

MOSASAURS:
THE GREAT SEA LIZARDS

THE LATE CRETACEOUS can be seen as the Age of the Mosasaur. These remarkable, often gigantic lizards dominated marine ecosystems right to the end of the Cretaceous, 66 million years ago. They included apex predators, blunt-toothed generalists, mollusc-eaters and slender-jawed specialists, and ranged from 30 cm (11¾ in) to 18 m (59 ft). As often seems the case in the history of life, mosasaurs owed their success to the downfall of other groups, their rise to dominance coming after the changes of the Cenomanian age resulted in the extinction of ichthyosaurs and pliosaurids.

Excellent skeletons, some with skin impressions preserved, show that 'classic' mosasaurs were powerful, gigantic predators with paddle- or wing-shaped limbs and a vertical tail fin. Individual genera tend to be widespread globally, as expected for animals capable of swimming vast distances. Once imagined to resemble giant crocodiles with paddles, we today think of mosasaurs as 'whale-lizards', highly streamlined and propelled by the end of the tail. Thousands of tiny scales covered the skin. Mosasaur teeth are approximately conical but prominent keels separate the inner and outer faces. The faces vary in terms of how flat or convex they are. Teeth are present on the rear of the palate in addition to the jaw edges. Joints in the lower jaw enhanced their ability to manipulate and swallow prey, and stomach contents confirm that large species were predators of other marine reptiles and large fish. Several early mosasaur lineages, however, consist of small, lightly built

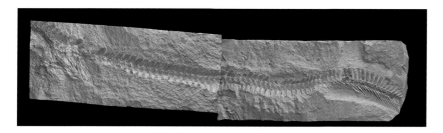
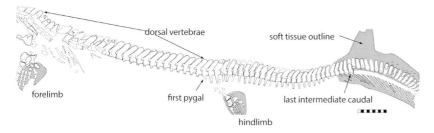

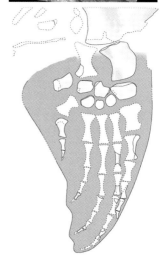

animals that exploited very different prey. These early kinds lacked the flipper-like limbs and specialized tails of classic forms and look like animals in the earliest stages of adapting to marine life.

Several different terms are used for the various mosasaur groups, and we need to explain these before moving on. Classic mosasaurs – the large forms that generally have flipper-like limbs and a large tail fin – are the mosasaurids, formally Mosasauridae. There are several subgroups within Mosasauridae. Mosasaurids descended from smaller animals called aigialosaurs. Together, aigialosaurs and mosasaurids are united as mosasauroids, formally Mosasauroidea. Finally, mosasauroids and another group – the dolichosaurs – are united within Mosasauria.

The consequence is that the term 'mosasaur' has several potential meanings. When most experts use it, they have in mind mosasaurids, yet the term mosasaur also applies to the more inclusive groups Mosasauroidea and Mosasauria. Throughout this book, 'mosasaur' is used for all members of Mosasauria. When members of the more specialized Mosasauridae are referred to, they are called mosasaurids.

MOSASAUR DIVERSITY

Mosasaurs are diverse in anatomy and size. They include small species that were not unlike terrestrial lizards, all the way to giants that were as specialized for life in the water as modern whales. Most mosasaurids were adapted for aquatic life. They lacked claws and their digits were incorporated into flippers. This is called the hydropedal condition. These fully aquatic forms also lacked a bony connection between the hip girdle and spine. This is called the hydropelvic condition.

A few exceptional fossils – like this *Prognathodon* from Jordan – reveal details of mosasaur skin anatomy as well as tail and limb shape. This fossil preserves a bilobed, shark-like tail (visible in the images at left) as well as flippers where the soft-tissue edge extends well beyond the internal bony anatomy.

Such animals would have been incapable of movement on land and, like whales today, would have died due to overheating and stress if stranded. The caveat here is that even fully aquatic animals *can* visit the land, but only when lunging onto the shore to grab prey that has come close to the water's edge. This behaviour is best known in killer whales, but is also practised by certain catfishes, eels and other fishes. We have no evidence that such behaviour occurred in mosasaurs and probably never will, but it remains a viable speculation that some mosasaurs captured terrestrial animals in this way.

Mosasaurs that lack flipper-like limbs (and instead have separated, clawed digits) are termed plesiopedal. They tend to have a bony connection between the hip girdle and spine. This is termed the plesiopelvic condition. As expected, the earliest mosasaurs – the dolichosaurs and aigialosaurs – are plesiopedal and plesiopelvic, this being the normal configuration widespread in lizards. Mosasaurids evolved from such forms, and it has been assumed that mosasaurids switched quickly from the ancestral configuration to the more specialized hydropedal and hydropelvic one.

A complication, however, is that some plesiopedal mosasaurs might not be early, archaic members of the mosasaurid group, since some studies find them to occupy a position in the family tree where they are surrounded by hydropedal species. It is not yet clear what this means. Perhaps the hydropedal condition evolved several times independently.

MOSASAURS IN THE LIZARD FAMILY TREE

Comparatively little argument has existed over the position of mosasaurs in the reptile family tree. They are definitely part of Squamata, the group that includes snakes and lizards. Exactly how mosasaurs are related to other

This simplified mosasaur family tree conveys the idea that mosasaurs began as small, amphibious, plesiopedal and plesiopelvic animals before developing aquatic specializations and gigantic size. If the tethysaurines really were plesiopedal (as shown here), it might be that the hydropedal condition evolved independently more than once.

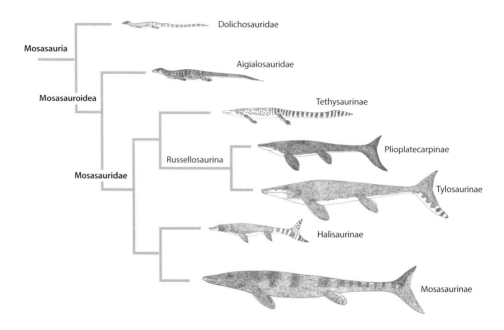

squamates remains controversial, however. The mosasaur skull looks like that of a monitor lizard, and it is mostly for this reason that mosasaurs have generally been regarded as part of Anguimorpha, the group that contains alligator lizards, slow-worms and gila monsters in addition to monitors.

When the mosasaur skull is compared in detail to that of monitors, however, it becomes less clear that mosasaurs and monitors are close relatives. The similarities are superficial, and probably the result of convergent evolution. Since the late 1990s, several experts have argued that mosasaurs are more closely related to snakes. They reason that the limb-bearing, marine snakes of the Cretaceous provide support for this connection, and in turn believe that snakes themselves originated as marine animals. The more traditional view of snake origins is that they originated as land-living burrowers.

This suggested link between mosasaurs and snakes adds further complexity, because DNA studies show that snakes are not anguimorphs but a squamate lineage all their own, albeit one that shared an ancestor with anguimorphs and with the group that includes iguanas. If, then, mosasaurs are related to snakes, they might not be anguimorphs after all. The possibility that mosasaurs might not be members of the snake or anguimorph lineages also exists, in which case mosasaurs evolved during the Jurassic from an ancestor that also gave rise to geckos, skinks and anguimorphs. This area is currently quite confusing, competing views exist among specialists, and new studies appear regularly.

> Mosasaurs are unquestionably lizards, and perhaps close relatives of such living groups as gila monsters (left) and monitors (right). Like those modern groups, mosasaurs probably started their evolutionary history with a forked tongue and a predatory lifestyle that involved the tracking and capture of vertebrate prey.

BIOLOGY AND BEHAVIOUR

Historically, our knowledge of mosasaur biology has focused on feeding behaviour and the fact that mosasaurids sometimes fought one another. More recently, insights into soft tissue anatomy, growth, reproduction and physiology have been made. Mosasaurid bone texture shows that they grew faster than expected for squamates, and both this and evidence from oxygen

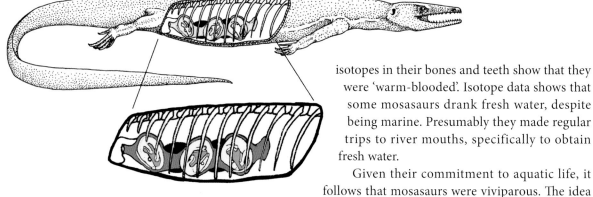

A *Carsosaurus* specimen described in 2001 reveals the presence of four embryos inside the body of the mother. The embryos have well-developed bones, suggesting that they were fully capable of swimming right after birth. All are oriented so that the tail would emerge first, as is common in aquatic animals that give live birth in water.

isotopes in their bones and teeth show that they were 'warm-blooded'. Isotope data shows that some mosasaurs drank fresh water, despite being marine. Presumably they made regular trips to river mouths, specifically to obtain fresh water.

Given their commitment to aquatic life, it follows that mosasaurs were viviparous. The idea that they might have crawled onto land to lay eggs – suggested in the late 1800s – is as likely as modern whales emerging onto land to give birth. The first good fossil evidence for mosasaur viviparity came from the discovery of abundant juvenile remains – including newborns – in the Mooreville Chalk of Alabama during the 1980s, and mosasaur viviparity has since been confirmed by additional specimens, some of which belong to large, hydropedal mosasaurids. Arguably more important is a specimen of the plesiopedal *Carsosaurus*, described by Canadian palaeontologist Michael Caldwell and Australian Mike Lee in 2001, that has at least four young within its body. This shows either that mosasaurs evolved viviparity early in their invasion of the seas, or inherited it from ancestors that were already viviparous, in which case it helped their transition to aquatic life. Viviparity evolved on more than 100 occasions in squamates and is present in numerous snakes and anguimorph lineages.

Potentially contradicting mosasaur viviparity is the 2020 discovery of a soft-shelled egg, 29 cm (11½ in) long, in the Upper Cretaceous sediments of Antarctica. Because it has features typical of squamate eggs, some experts suggest that it might belong to a mosasaur. While not impossible, this is unlikely and it was probably laid by another reptile, perhaps a pterosaur.

For a time, it was thought that mosasaurs gave birth in rivers, estuaries or lagoons where the young could avoid the predators of the open ocean. However, finds made in Kansas and elsewhere show that mosasaur juveniles lived in the pelagic realm, far from land, and there is no evidence that mothers gave birth in special, protected places. We assume that juveniles were independent and lived separately from their mothers.

If mosasaurs are allied to anguimorphs or snakes, they likely had a forked tongue, and used it in detecting chemical particles in the water as aquatic snakes and lizards do today. No evidence for the shape of the mosasaur tongue has been discovered, so we remain uncertain whether they had a muscular tongue with a short, forked region (like that of gila monsters) or a more slender one with a long, forked region (like that of monitors and snakes).

Another feature common to anguimorphs and snakes might have been present in mosasaurs: venom. Venoms are not unique to snakes and the gila monster family as once thought, but widespread in anguimorphs and the iguana group, too. These venoms are not stored in large venom sacs and injected via syringe-like fangs, as is the case in vipers and cobras, but kept in long glands present along the length of the lower jaw and connected to the outside via ducts between the teeth. It could be that mosasaurs had venom and used it to weaken or kill prey, but it is also possible that they lost it over time as it became less useful. After all, releasing venom underwater could be wasteful and ineffective.

The fact that mosasaurs are squamates also means that traditional reconstructions of their faces are inaccurate. Mosasaurs are usually shown as crocodile-like, with prominent teeth and tightly fitting facial skin. Squamates are different from this, since their teeth are largely hidden by gum tissue and large, scale-covered lips conceal the gums and jaw edges from outside view. It should be assumed that mosasaurs were like this, too, and such is confirmed by a Russian specimen that has impressions of its facial soft tissues preserved.

We remain ignorant of mosasaur social behaviour. The stereotypical view is that they were solitary and fought regularly with members of their own species. This could be true. However, snakes and lizards today are sometimes social, sometimes forage in groups and sometimes live together in mated pairs or families. It is plausible that some mosasaurs also behaved in this way.

DOLICHOSAURS AND AIGIALOSAURS

Two plesiopedal Cretaceous groups – dolichosaurs and aigialosaurs – differ from mosasaurids in being considerably smaller, and in having long, slender, flexible bodies and tails. Their skulls and tails show that they were swimmers, but how aquatic they were is disputed. The conventional view is that they also foraged on land. Several details, however, suggest that they were highly aquatic. They are mostly associated with the Western Tethys Ocean, but their fossils are widespread, so they travelled far and wide during their history.

Dolichosaurs were 30 to 100 cm (11¾ to 39½ in) long and would have resembled limbed, chunky-bodied snakes or long-bodied, skinny monitors. They might be early members of Mosasauria (and thus be ancestral to aigialosaurs), or they could be close relatives of snakes and lack any close relationship with mosasauroids. *Coniasaurus* from the Late Cretaceous of England, Germany and the USA is among the best known of the dolichosaurs. Its long, shallow snout, closely packed, bulbous teeth and long neck suggest specialized feeding habits. Perhaps it foraged among reefs in search of crustaceans and fishes. *Aphanizocnemus* from Lebanon is known from a skeleton about 30 cm (11¾ in) long. It shows that some dolichosaurs were especially long tailed, with reduced limb girdles, and limbs where some of the bones are unusually short. The claws are small and straight, rather than curved. These features point to an aquatic existence. The Italian *Primitivus* shows that dolichosaurs survived almost to the end of the Cretaceous.

Aigialosaurs have been known since 1892 when *Aigialosaurus* was described from the Upper Cretaceous of Croatia, specifically from Cenomanian or Turonian sediments dating to between 100 and 90 million years ago. Similar animals were later described from Slovenia, Israel, the USA and Mexico. All are about the same geological age. Aigialosaurs were mostly 1 to 2 m (3¼ to 6½ ft) long and their skulls superficially resemble those of monitors. The skull is narrow, shallow and lightly built, the teeth are small, conical and widely spaced, and a joint half-way along the lower jaw provided flexibility during feeding.

Aigialosaur limbs are not obviously modified relative to those of terrestrial lizards. However, the aigialosaur pelvis is reduced and smaller than normal for terrestrial lizards. The tail is long (sometimes with more than 75 vertebrae) and the bony spines projecting from its upper and lower margins are long, slender and inclined towards the tail's tip. The sideways-projecting bony bars termed transverse processes are present at the tail's base but are lacking from most of its length, in contrast to the set-up typical for lizards. This was a flexible, narrow tail that would have formed a sculling organ, and a specimen of the Mexican *Vallecillosaurus* shows that a low, horizontal fin was present along its length. *Vallecillosaurus* preserves 15 tiny pebbles in its stomach region. Modern lizards sometimes swallow small stones for use in digestion, though this is only known in omnivorous or herbivorous species. Maybe this shows that *Vallecillosaurus* was more flexible in diet than other aigialosaurs, or maybe aigialosaurs in general were not as predatory as assumed. Several aigialosaur fossils have patches of skin preserved. Their scales are rhomboidal, smooth, and arranged in oblique rows. Similar scales are known for dolichosaurs. In addition, the dolichosaurs *Pontosaurus* and *Primitivus* preserve broad scales on the underside of the tail, similar to those on the underside of snakes. Aigialosaurs share a list of features with mosasaurids, such as the same kind of flexible joint mid-way along the lower jaw, but lack features that tie them together as a discrete group of related species. What seems likely is that they therefore include the direct ancestors of mosasaurids. The fact that they are mostly older, smaller and less specialized than the oldest mosasaurids is consistent with this view.

TETHYSAURINES

Around 100 million years ago, members of one aigialosaur lineage evolved larger size and flipper-like limbs, and became better suited for aquatic life. They gave rise to the mosasaurids, a group that quickly diversified into five main subgroups. Mosasaurid evolution was complex and there are suggestions that certain aquatic features present in these groups – such as hydropedal limbs – evolved more than once.

One poorly known mosasaurid subgroup is known from North America, countries in Africa and eastern Europe. All its members are twenty-first century discoveries, and indeed the group itself was only named in 2012. This is Tethysaurinae, named after *Tethysaurus* from the Late Cretaceous of Morocco. *Tethysaurus* was around 3 m (10 ft) long, and is known from complete, articulated skeletons. Several closely related animals are known, including *Yaguarasaurus* from Colombia, *Russellosaurus* from Texas and *Romeosaurus* from Italy. Tethysaurines have longer tail vertebrae and a narrower shoulder blade than is typical for mosasaurids, and none appear to have exceeded 6 m (19¾ ft).

The most newsworthy thing about this group is that it includes a freshwater member. This is *Pannoniasaurus* from Hungary, the fossils of which were preserved in an environment where lakes and river channels were surrounded by swamps and woodlands inhabited by dinosaurs. The skull is flattened and crocodile-like and it seems that it was an ambush predator that grabbed prey from the water's edge. Even more surprising is that *Pannoniasaurus* is plesiopelvic. The suggestion has been made that it was plesiopedal, but this has yet to be established with confidence since its limbs are incompletely known. Ordinarily, we would assume that the presence of these configurations in a mosasaurid would be due to it being close to the group's origin, and still in possession of features that it retained from aigialosaur-type ancestors. This might be the case.

However, some experts argue that tethysaurines are not close to mosasaurid origins, but are instead surrounded on the family tree by fully aquatic, hydropedal mosasaurids. If this is accurate, several possibilities exist. One is that hydropedal mosasaurid groups evolved hydropedal limbs independently of one another. Some experts argue that this is plausible, since these groups differ more from one another than thought. Another possibility is that *Pannoniasaurus* (and maybe all tethysaurines) evolved from hydropedal ancestors but reverted to a plesiopedal configuration due to specialization for an unusual lifestyle.

The final possibility is that the describers of *Pannoniasaurus* are wrong, and that it was hydropedal like other mosasaurids. Only further finds will resolve this mystery.

> The Moroccan mosasaur *Tethysaurus* is known from several specimens, most of which preserve much of the vertebral column. When first published in 2003, *Tethysaurus* was argued to be an evolutionary intermediate between aigialosaurs and later mosasaurs. The specimen shown here has yet to be the focus of scientific research.

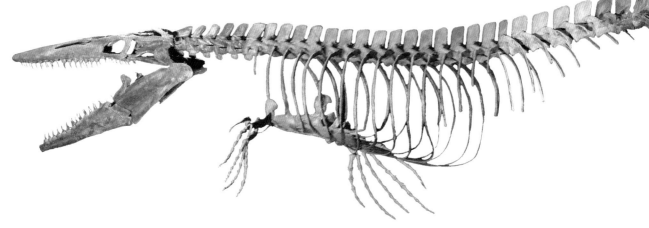

TYLOSAURINES

Few mosasaurids have been depicted in art and popular culture more frequently than *Tylosaurus*, a widespread, geologically long-lived animal known from numerous specimens. *Tylosaurus* was first recognized in 1869 when American palaeontologist Edward Cope described an enormous partial skull – more than 1.5 m (5 ft) long when complete – from the Upper Cretaceous of Kansas. *Tylosaurus* and its kin – the tylosaurines – were mostly around 10 million years older than *Mosasaurus* and its kin, and more archaic overall. However, they did persist to the end of the Cretaceous.

Tylosaurines share features with plioplatecarpines, and both are allied within the clade Russellosaurina, the name of which commemorates the studies of Dale Russell. The russellosaurine radiation is a different branch of the family tree to the mosasaurine branch. Both evolved in parallel, both gave rise to specialized species during the later parts of their history, and both persisted to the end of the Cretaceous.

Tylosaurus was hydropedal, hydropelvic, and enormous. The biggest specimens exceeded 14 m (46 ft), making them among the biggest of marine reptiles. These giants aren't typical for the genus, however, but belong to the youngest of the *Tylosaurus* species: *T. proriger*, the one named by Cope for that partial skull from Kansas. Older *Tylosaurus* species were smaller, mostly less than 10 m (32¾ ft). This trend in increasing size is known as Cope's Rule. *T. proriger* also has features that make it look paedomorphic (see Chapter 3) relative to older, smaller *Tylosaurus* species. It might be both a specialized giant and an animal that retained juvenile features into adulthood.

Tylosaurus has a long, shallow, pointed snout with an especially streamlined appearance. A deep bulge half-way along the lower jaw's lower edge helps create its distinct profile, as does a bony tip to the snout, termed the prow, which extends ahead of the teeth. The prow probably played a role in striking or stunning other animals. Like the bony beak of certain dolphins, perhaps it was used as a ram, driven at speed into a prey animal or opponent to kill or incapacitate.

Stomach contents show that *Tylosaurus* ate other mosasaurs, plesiosaurs, turtles, seabirds, bony fishes and sharks. Bite marks attributed to *Tylosaurus* have been reported from ammonite shells, turtle shells and dinosaur bones as well as other *Tylosaurus* specimens. Evidence for *Tylosaurus* versus *Tylosaurus* violence is known from several specimens, mostly from the jaws and snout, and there are cases where the attacks appear to have been fatal. Maybe they fought for territorial reasons, because they sometimes engaged in cannibalism, or because larger individuals sometimes killed smaller ones when they could.

Tylosaurine flippers are different from those of most other mosasaurid groups. Their forelimb bones are longer and not as compressed and squat as those of other mosasaurids, and the hand is long and wide, the arrangement of digits suggesting a broad, fan-like shape. Exactly what this means for swimming behaviour isn't clear, but studies are underway.

This spectacular, 12 m (40 ft) *Tylosaurus* specimen – nicknamed the 'Bunker mosasaur' after finder C D Bunker – was discovered in Kansas in 1911. It is one of the largest mosasaur specimens ever discovered. It reveals the long, streamlined snout, bulging lower edge to the lower jaw and broad, fan-like limbs of this mosasaur group.

HALISAURINES

Halisaurinae is a third mosasaurid subgroup, named after *Halisaurus* from New Jersey, USA. Halisaurines are known from Sweden, Belgium, Morocco, Niger, Nigeria, Colombia and Japan, and members of the group were in existence for the last 20, and probably 30, million years of the Cretaceous.

In several aspects, halisaurines appear unspecialized relative to mosasaurid subgroups such as tylosaurines and mosasaurines. The halisaurine spine, for example, is more flexible than that of those other mosasaurid groups. Halisaurine limbs are not as flipper-like as those of more familiar mosasaurids, since the humerus, radius and ulna are long and the number of phalanges is relatively low. A traditional view was therefore that halisaurines are primitive mosasaurids. This is contradicted, however, by studies that find halisaurines to be closer to mosasaurines than to mosasaurids such as tylosaurines. This is again suggestive of complex evolution within the group, and with the view that tylosaurines and mosasaurines evolved in parallel from less specialized ancestors.

The anatomy of halisaurines such as *Halisaurus* and *Eonatator* from the USA, Sweden and Colombia suggests that these animals were generalist predators that used their small, delicate, pointed teeth to grab fish and squid. Most halisaurines were probably like this, but a few were unusual. *Phosphorosaurus* from Belgium and Japan has especially big eyes and a skull shape suggestive of binocular vision. Perhaps it was good at hunting in low light, either at night or at depth. Meanwhile, the several *Pluridens* species

Mosasaurs like the halisaurine *Pluridens* may well have used the tongue to collect scent particles when hunting. *Pluridens* has proportionally small eyes for a mosasaur, so perhaps it used scent and its sense of touch more than other members of the group.

from northern and western Africa have especially long, slender jaws and high tooth counts. They superficially resemble various Jurassic ichthyosaurs, and they presumably took snapping bites at small, swift prey. Halisaurines were mostly 3 to 4 m (9¾ to 13 ft) long or less, but some *Pluridens* species reached 8 or 9 m (26 or 29½ ft).

Pluridens was part of a Late Cretaceous African community which included numerous species belonging to several mosasaurid lineages. This shows that mosasaurid diversity was high right to the end of the Cretaceous, and that some marine environments at the time were inhabited by 10 or more mosasaur species.

PLIOPLATECARPINES

Plioplatecarpines are a hydropedal mosasaurid subgroup increasingly regarded as one of the most successful, long-lived and widespread mosasaurid groups. Plioplatecarpines were mostly small to mid-sized, from 3.5 to 7.5 m (11½ to 24½ ft), compact, short-snouted, and with proportions indicating agility. The namesake of the group – *Plioplatecarpus* – was described from Belgium by Louis Dollo in 1882, but later discovered in

The skull of *Platecarpus* is proportionally short-snouted for a mosasaur. The slender teeth suggest a diet of small fish and squid rather than other marine reptiles. Features to note here include the joint mid-way along the lower jaw and the sclerotic ring within the eye socket (a typical feature of the reptilian skull).

Sweden, the Netherlands, the USA and Canada. Large eyes and forward-angled teeth suggest specialized hunting behaviour.

One of the most familiar plioplatecarpines is *Platecarpus*, described by Cope from the Kansas chalk in 1869. It is mid-sized, around 6 m (19¾ ft) long and short-snouted for a mosasaur, with a low tooth count, there being only 11 teeth in each half of the upper jaw and 10 in each half of the lower jaw. *Platecarpus* is especially important for our knowledge of mosasaur biology thanks to a specimen preserved with skin, parts of the windpipe and even remnants of the eyeball and internal organs. The specimen also preserves a large triangular upper tail lobe.

A radical view of *Plioplatecarpus* was published by mosasaur expert Theagarten Lingham-Soliar in 1992. He pointed to its long, wing-like forelimbs, deep chest and muscular upper arm and pectoral girdle as evidence that it was doing something not previously suggested for any mosasaur, namely underwater flight. Lingham-Soliar used *Plioplatecarpus* to help make the point that the mosasaurids of the very latest Cretaceous were extremely diverse. This view of high diversity has since received support from other discoveries, but Lingham-Soliar's view of a penguin-like *Plioplatecarpus* has not proved correct. Its flippers were broad, not wing-like, and it did not have the shoulder anatomy required for underwater flight.

Plioplatecarpus might not have been as unusual as once argued, but it remains the case that other plioplatecarpines are strange. *Ectenosaurus* from the USA has a stretched-looking, narrow snout and high tooth count, with 17 teeth in each maxilla, a place where most mosasaurs have 12. It was likely a specialist predator, perhaps of fast-moving fish.

Plioplatecarpus – namesake member of the group Plioplatecarpinae – is even shorter-snouted than *Platecarpus*. Especially big eyes are also typical of this animal, as are teeth with recurved crowns. *Plioplatecarpus* was widespread and several species have been named.

We've known since the 1960s that plioplatecarpines inhabited the seas of northern and eastern Africa. The small *Angolasaurus* was named for its discovery in Angola, later finds demonstrating its presence in Niger, and North and South America. The related *Gavialimimus* from Morocco had unusually long jaws, and this explains its name, meaning 'gavial mimic'. The seas covering Morocco and nearby were a mosasaur diversity hot-spot and numerous species lived here, apparently because this was an exceptionally fertile area where upwelling currents met the coast. *Selmasaurus* from the USA, another member of the *Angolasaurus* group, is unusual because it lacks cranial kinesis (see Chapter 3) while *Goronyosaurus* from Niger has a blunt snout and teeth that interlocked tightly as the jaws were closed.

Why plioplatecarpines tended to be unusual relative to other mosasaurids is unclear. One possibility is that they were filling niches unexploited by other marine reptiles of the time. Their diversity shows that mosasaurids were evolving quickly within the last few million years of the Late Cretaceous and that new species and even new lineages were evolving. They were not a stagnant group heading for extinction but the very opposite.

MOSASAURUS AND KIN

Finally, we come to the mosasaurines, the group that contains *Mosasaurus* itself. Not only does *Mosasaurus* have an important position in the history of Mesozoic marine reptile research (see pp. 16–17) it is also one of the largest mosasaurs and largest of marine reptiles. The biggest individuals reached 15 m (49 ft), with 17 or 18 m (55¾ or 59 ft) estimated by some experts. Key mosasaurine features mostly involve their forelimbs, since they possess arm and hand details that relate to flattening and stiffening of the limb.

Close relatives of *Mosasaurus* include *Plesiotylosaurus* and *Plotosaurus* from California and *Eremiasaurus* from Morocco. These are mostly from the last part of the Late Cretaceous and their diversity is in keeping with the idea that mosasaurid evolution was occurring rapidly right up to the extinction event. *Plesiotylosaurus* has a slender snout that resembles that of *Tylosaurus*. *Plotosaurus* had sharp keels on its teeth, a deep, stiffened body and was more reliant on propulsion from the fluked part of its tail than other mosasaurids. It was perhaps the most ichthyosaur-like or whale-like of all mosasaurids, and the suggestion has been made that it would have given rise to increasingly streamlined, pelagic descendants had it not gone extinct.

The group also includes *Clidastes*, which was around 6 m (19¾ ft) long. The name *Clidastes* was first given to *C. propython* from the USA (a species now also known from Sweden) but remains discovered worldwide were later named as new species within the same genus. Today it is thought that only a few – perhaps none – of these other species are closely related to *C. propython* and it is expected that future work will pin their affinities down and give them new names. *Clidastes* is positioned close to the origin of mosasaurines and it is suspected that it was ancestral to later members of the group.

Also among the mosasaurines are around 10 mid-sized to gigantic, chunky-headed species, some of which possess forward-slanted teeth at the front of the upper jaw. These are the *Prognathodon* species, known from Belgium, the Netherlands, Spain, Ukraine, Israel, Jordan, Angola, the USA, Canada and New Zealand. Substantial variation is present across these animals, and they are another group where the currently used nomenclature fails to reflect our understanding of their evolutionary relationships. There are probably four or five genera here, widely separated on the family tree. *P. currii* from Israel probably reached 11 m (36 ft). Its thick, deep jaws suggest a powerful, crushing bite.

Some *Prognathodon* species are closely related to *Globidens*, first described from the USA but known from Morocco and Angola too. *Globidens* was 6 m (19¾ ft) long and equipped with a heavy lower jaw and spherical teeth. Experts have often suggested that these demonstrate specialization for a diet of shellfish, involving either giant clams or ammonites. Stomach contents from a specimen found in South Dakota confirm that *Globidens* consumed clams. However, living lizards with rounded teeth are generalists that consume a variety of prey, and it might be that *Globidens* was a generalist too.

Globidens was not unique in possessing rounded teeth, since they were also present in the smaller (3.5 m, 11½ ft, long) *Carinodens* of Europe and Africa, and *Igdamanosaurus* of northern Africa. Teeth of an entirely different sort are present in the small *Xenodens* from Morocco, which was less than 2 m (6½ ft) long. These have hooked, blade-like crowns and are low and closely spaced. They might have served a cutting function, perhaps used on the bodies of much larger animals. No other Mesozoic marine reptile has teeth of this sort, and here again is an animal from the very end of Cretaceous times, the final days of mosasaurian success.

This skeletal reconstruction of *Clidastes* (with human for scale) emphasizes the broad, rounded flippers of this animal. This is specifically *C. propython*, a species named by Edward Cope in 1869 for a specimen from Alabama, USA.

Rounded teeth evolved several times within mosasaurs. This partial lower jaw probably belongs to *Globidens*. As is typical for marine reptile fossils from the Upper Cretaceous sediments of Morocco, the bone is white, rather than the brown or black more commonly seen in fossil bones.

9 SEA TURTLES

This diagram of a living sea turtle's skeleton omits the plastron – that part of the shell that covers the underside – for ease of viewing. It is obvious that the main organs of propulsion are the enormous forelimbs.

TURTLE IS THE CATCH-ALL name we use for any member of the reptile group Testudines. Turtles originated during the Triassic as small, armoured animals of inland environments, but not until the Jurassic did certain groups take to living in ponds, lakes and rivers, and eventually the sea. Here, they evolved into pelagic giants. The evolutionary history of turtles is long and complicated, but we will focus on the sea turtles here, as they are, of course, part of the Mesozoic marine reptile story.

Turtles have a unique body plan. The ribcage is fused with a series of new bones to create the shell, the upper part of which is the carapace and the lower part the plastron. The limb girdles are thus within the ribcage. Horny plates termed scutes or scales usually cover the shell. All turtles except a few early forms are toothless, a horn-covered beak functioning in the capture of prey, cropping of vegetation, or crushing of shells. Despite the limitation posed by the shell, turtles have given rise to amphibious omnivores and predators, terrestrial herbivores, and freshwater and marine forms.

The turtles alive today that live in the sea belong to Chelonioidea, the history of which extends back to the middle of the Cretaceous, around 115 million years ago. Two main groups belong here: the hard-shelled sea turtles and the leatherbacks. A third group – the protostegids – is extinct and restricted to the

Cretaceous. Views differ on whether protostegids are chelonioids or not and several possible evolutionary positions have been suggested. The most traditional idea is that protostegids are chelonioids and might have a close affinity with leatherbacks. Since about 2007, experts have increasingly favoured the idea that protostegids are outside the group that contains living chelonioids, and some experts even think that protostegids lack any close connection with chelonioids at all.

Chelonioid beginnings are obscure, but they likely evolved from a Cretaceous and probably North American ancestor that also gave rise to the swamp-dwelling snapping turtles and mud turtles. Presumably, this was a mid-sized turtle of estuaries that took to swimming at sea and eventually evolved flippers, salt glands and other marine adaptations.

Members of other turtle groups – including the side-necked turtles – also include fossil sea-going species. Side-necked turtles today are overwhelmingly associated with freshwater environments. It is partly for this reason that experts largely missed the fact that numerous fossil side-necked turtles – including many from the Cretaceous – were marine. Several turtle groups of the Jurassic – the plesiochelyids, thalassemydids and eurysternids, all of which are included within the group Thalassochelydia – also include marine species. Some of these were large, with carapace lengths of 1 m (3¼ ft), had fully webbed paddles or flippers, and inhabited shallow marine and coastal environments. Their fossils come from the Solnhofen region of Germany as well as Switzerland, France, England, Spain and Portugal. They show that sea-going turtles were typical of Jurassic Tethyan environments. Protostegids might have their origins within thalassochelydians rather than chelonioids.

The underside of *Solnhofia*, a eurysternid sea turtle of Jurassic Europe. The bones of the plastron have a diamond-like configuration. That hole in the middle – termed the fontanelle – is not a typical turtle feature, but one that evolved repeatedly in groups with a reduced shell.

PROTOSTEGIDS

Among the most impressive turtles of all are the protostegids, a group first recognized by Edward Cope in 1871 for *Protostega* from the Upper Cretaceous chalk of Kansas. This was a gigantic turtle with huge, broad flippers and a broad, shallow shell 3 m (10 ft) long. Large spaces exist between the ribs, this being a lightening adaptation linked to pelagic life. The plastron was also reduced, a massive space existing in its middle.

An even larger close relative was discovered in South Dakota during the 1890s. *Archelon* was a giant 4.6 m (15 ft) in length, with a flipper-span

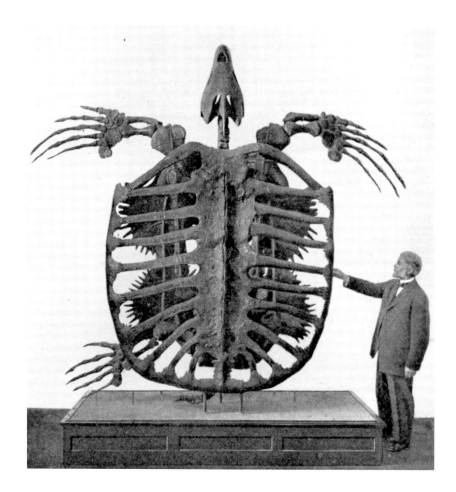

Archelon was gigantic, as is obvious from this photo. Normally in sea turtles, the ribs are rectangular and form a complete bony covering over the top of the skeleton, but in protostegids like Archelon they are reduced and there are long gaps between them. The serrated edges on the bones of the plastron are also obvious.

of 4 m (13 ft), again with a broad, flat, reduced shell. *Archelon*'s beak is long and hooked, whereas in most other protostegids it was shorter and only slightly curved. Thanks to its size, near-complete remains and short name, *Archelon* is the most famous fossil turtle of all, and also the only one to star in a Hollywood movie, *One Million Years B.C.* from 1966. The animal even gets a namecheck from one of the film's anachronistic cavepeople. Numerous additional protostegids have been discovered since. Most are European but some are from Russia, the Americas, Australia, New Zealand and Japan. Most are from the Late Cretaceous but the oldest is from the Early Cretaceous of Colombia.

We remain unsure how protostegids made a living. Stomach contents are unknown and they lack features that link them to any specific lifestyle. The long, hooked rostrum of *Archelon* shows that it was predatory. The possibility exists that it was an ammonite predator that used its rostrum to reach into ammonite shells and extract the soft body. A remarkably odd skull was present in *Ocepechelon* from the Late Cretaceous of Morocco. The skull alone is 70 cm (27½ in) long and has an elongate, narrow snout that is concave on its underside and open at the tip. Presumably, *Ocepechelon* used suction to draw prey into the mouth. It's difficult to imagine that this strategy would work while swimming through open water, where potential prey would be spread out, so perhaps *Ocepechelon* stuck its snout into cavities and crevices. The possibly related *Alienochelys*, also from the Late Cretaceous of Morocco, has a massively built, broad, thickened upper jaw and palate. This looks suited for the crushing of molluscs, crabs and other prey.

The precise anatomy of the protostegid shell is uncertain. Some protostegid bones bear markings left by large scales, but most have a reduced carapace and appear to have had a leathery covering like that of the leatherback. Maybe they had scales in some regions and a leathery covering elsewhere, or maybe only early species had scales.

The fact that protostegids lived alongside mosasaurs and pliosaurids means that they must have had to evade or fight off such predators. Maybe they fought back with flipper-strikes or bites, or maybe they used the width or depth of their large shells to prevent or obstruct an attack. Missing flippers and bite marks on shells demonstrate that encounters of this sort did happen, and protostegid bones have been discovered as mosasaur stomach contents. Bands of dead bone in the limbs of *Protostega* and some other protostegids have been interpreted as evidence for decompression syndrome, when the animal has risen too quickly from deep waters, causing nitrogen to form bubbles in the blood and tissues. This could show that some protostegids were deep divers, and perhaps used this behaviour to escape predators or find food.

We assume that protostegids were similar in other aspects of lifestyle to living sea turtles. Long, powerful fore flippers suggest fast, powerful swimming, and it might be that they were especially good at turning and

The large skull of an adult *Archelon* is around 70 cm (2¼ ft) long. The strongly hooked shape of the upper jaw is obvious and suggests a predatory habit. In life, a thick horny beak would have covered both the upper and lower jaws.

The huge skull of *Ocepechelon* – shown here in side view, facing right – is one of the strangest turtle skulls of all. The nostril openings are located on the forehead, in between the eye sockets.

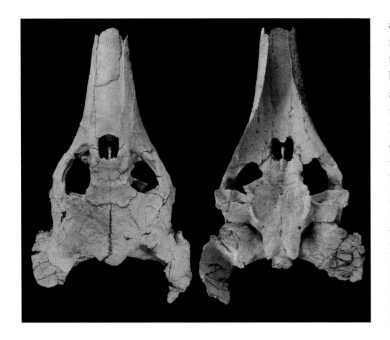

These images of the *Ocepechelon* skull show it as seen from above (left) and from below. The narrow rostrum of this unusual turtle is concave on its underside.

accelerating. They were almost certainly identical to living sea turtles in that they laid eggs on land, the consequence being that females must have made arduous treks along beaches to excavate their nests. Fossil turtle eggs from the Late Cretaceous of Japan might be those of protostegids, and a Late Cretaceous turtle nest from the USA might have been made by a protostegid – it looks much like a nest made by a modern sea turtle. The hatching of protostegid eggs was perhaps similarly timed to match high tides, and this presumably resulted in the mass emergence of numerous hatchlings that had to make their way to the water before they were picked off by dinosaurs and other predators.

The V-shaped mass of preserved sediment visible in the middle of this image appears to represent a fossilized sea turtle nest. Unlike most other marine reptile groups, turtles never – so far as we know – abandoned the ancestral egg-laying habit typical for reptiles.

HARD-SHELLED SEA TURTLES

Living alongside the protostegids were members of a second sea turtle group. These are the hard-shelled sea turtles or cheloniids, six species of which exist today: the green, loggerhead, Kemp's ridley, olive ridley, hawksbill and flatback sea turtle. These are a diverse lot. Some are herbivores, some omnivores, and others mostly eat molluscs and other animals. Hawksbills and some Kemp's ridleys consume sponges, which is remarkable because sponges are mostly formed of silica, a material we associate with minerals and glass. Species such as the green sea turtle change their diet as they mature, becoming increasingly herbivorous.

Fossils and genetic studies show that the lineages leading to these living species split from one another within the past 20 million years. They therefore descend from an ancestor that lived during the later part of the Cenozoic, meaning that those hard-shelled sea turtles known from the Cretaceous were not members of the modern cheloniid group. Instead, these fossils are early members of the cheloniid lineage. Or, rather, *some* are: it now seems that the hard-shelled sea turtles of the Cretaceous belong to different parts of the family tree, and that sea turtle history in the Cretaceous was complicated, involving groups similar to cheloniids but distinct from them.

Cretaceous hard-shelled sea turtles were mostly small, with carapace lengths of less than 70 cm (27½ in). They had the long, broad flippers typical of modern sea turtles, most had a broad, flattened carapace, and short, robust jaws. A few, including *Prionochelys* from Kansas and Alabama in the USA, possessed sharp peaks along the carapace midline, the result being a saw-backed appearance. This looks like a defensive adaptation, used to deter mosasaur or plesiosaur attacks. The possibility that it was used in fights or mating displays, however, also has to be considered. *Prionochelys* belongs to a North American group called the ctenochelyids, species of which survived to the end of the Cretaceous.

Allopleuron from Belgium and the Netherlands was a different sort of Late Cretaceous sea turtle, some specimens of which have a carapace length of 1.5 m (5 ft). *Allopleuron*'s carapace was long, streamlined, and with reduced ribs. Scales were absent – making it superficially similar to a modern leatherback – so it was not hard-shelled at all. In fact, some experts regard it as an early member of the leatherback group. Its pointed, triangular skull suggests that it was grabbing fish.

Finally, *Toxochelys* from the Western Interior Sea is unusual in having partly upward-facing eyes and broad flippers, where some of the fingers

Toxochelys from the USA was not as specialized for swimming as were such sea turtles as protostegids. This skeleton is on display at the Houston Museum of Natural Science, USA.

were individually mobile. It was also large, reaching 2 m (6½ ft) in total. Though traditionally regarded as early members of the cheloniid group, *Toxochelys* and other similar forms are archaic enough that they might belong elsewhere on the turtle family tree, perhaps outside the group that contains all other sea turtles.

LEATHERBACK SEA TURTLES

A third sea-going turtle group emerged in the Cretaceous, survived the KPg Event and persists to the modern day. These are the leatherbacks, formally known as dermochelyids. A single species survives: *Dermochelys coriacea*, typically just called the leatherback sea turtle. It occurs globally, from the tropics to cold regions around Scandinavia and Alaska. Sadly, it is endangered and declining, the causes including disturbance at nesting beaches, nest exploitation by people, plastic pollution and collision with boats. The leatherback is a remarkable reptile, adapted for pelagic life. A thick fat layer keeps it insulated and it is able to dive to 1,200 m (3,937 ft) and forage in freezing waters. Leatherbacks grab sea jellies and similar animals using hooks and notches on their upper beak. Hundreds of fleshy spikes inside the mouth and throat help trap prey. However, they appear opportunistic, and also eat fish, molluscs, crustaceans and algae. Colossal front flippers, which can give large individuals a span of 2.7 m (9 ft), are used to generate acceleration and also serve as weapons in defence against sharks. The biggest individuals reach 2 m (6½ ft) in length and 700 kg (1,543 lbs) or more.

A few features make leatherbacks unusual relative to other sea turtle groups. Their shell is so reduced that they can be described as lacking one. The ribs are not merely separated by gaps but are reduced to tapering bars that almost fail to reach the carapace edges. The carapace itself is formed of thick, leathery skin and small bony plates. Seven parallel ridges arranged along the carapace might help channel water along the body. Leatherbacks have a high internal temperature and confusion exists over whether they are 'warm-blooded' or not. The latest studies show that they generate heat internally, maintain high internal temperatures, and have adaptations that slow or prevent this heat escaping to the outside, so they are indeed 'warm-blooded' or endothermic.

The oldest fossil leatherbacks are from the Late Cretaceous. *Turgaiscapha* from Kazakhstan and *Corsochelys* from the USA have been identified as

> The modern leatherback sea turtle is virtually always blue mottled with white, but some individuals are almost black. Living leatherbacks appear divided into four distinct populations, the most abundant of which is the Atlantic one.

Late Cretaceous leatherbacks, but this is controversial and better fossils are needed to confirm it. Leatherbacks are similar to protostegids in some respects and some experts have argued that they are closely related. If this is correct, leatherbacks should have a fossil record that extends back to the Early Cretaceous given that the oldest protostegids come from this time. This would mean that we lack the first 50 million years of leatherback history.

The best-known Cretaceous leatherback is *Mesodermochelys* from Japan, numerous specimens of which are known. It reached sizes greater than those of the living leatherback, since some specimens have a carapace length of 2 m (6½ ft). It would also have looked different from *Dermochelys* since its carapace was not as reduced. The bones forming the edges of its carapace, known as peripheral bones, are thick, and its ribs are rectangular and in contact for most of their length. Its skull is shorter and broader than that of *Dermochelys*. *Mesodermochelys* is known from numerous sites, many of which also yield mosasaur and plesiosaur remains. A peculiarity is that hard-shelled sea turtles are mostly unknown from the places that yield Cretaceous leatherbacks, so some experts argue that sea turtle distribution in the Cretaceous was what's known as provincial: that is, that the different groups were restricted to particular areas. Leatherbacks were abundant in the West Pacific while hard-shelled sea turtles were not, and the reverse was true in the North Atlantic.

Why and how leatherbacks survived the KPg Event while protostegids did not is unclear. The anatomy of *Mesodermochelys* and behaviour of *Dermochelys* suggest that Late Cretaceous leatherbacks were generalists, able to feed on algae and a variety of bottom-dwelling invertebrates, resources that were still findable during a time when marine ecosystems were disrupted.

It would be tragic if, after having survived so long, and made it through times of such hardship, these and other sea turtle groups were killed off by human impact. There was a time decades ago when people entertained the possibility that a warmer future might result in conditions suitable for the evolution of giant marine reptiles again, that small lizards might give rise to 'neo-mosasaurs' and that great marine serpents might descend from the sea snakes of today. The fact is that Earth faces an impoverished future, where oceanic vertebrate diversity is greatly diminished, and where the global oceans are left in a state not that different from the one that followed the end-Permian event.

A *Mesodermochelys* skeleton on display at Hobetsu Museum in Mukawa, Japan. The bones of the carapace are larger and thicker than those of *Dermochelys*. The edges of its jaws are also thicker than those of *Dermochelys*, perhaps meaning that it consumed harder foods.

GLOSSARY

Anatomy Anything pertaining to the way organisms are put together and how their parts and structures function. The term anatomy can refer both to the structure and function of these parts, and to the science of studying them.

Archosauromorphs The diapsid reptile clade that includes archosaurs and several related groups, including protorosaurs. Archosauromorphs originated in the Permian and are still alive today. The marine reptile superclade might be part of, or close to, Archosauromorpha.

Archosaurs The archosauromorph clade that includes crocodylomorphs and all of their relatives, and dinosaurs and all of their relatives. The name means 'ruling reptiles', a reference to archosaur prominence in Mesozoic ecosystems.

Cenozoic The great section of time – the so-called 'Age of Mammals' or 'age of new life' – that began 66 million years ago and continues today. Birds have outnumbered mammals throughout the Cenozoic, so it might be better named the 'Age of Birds'.

Clade A group of organisms where all the included populations descend from the same single population.

Cretaceous The section of the Mesozoic that lasted between 145 and 66 million years ago. Laurasia and Gondwana began to break apart to form the modern continents during the Cretaceous. Many environments would have looked somewhat modern.

Crocodylians The crocodylomorph clade that includes crocodiles, alligators and gharials. Crocodylians are the only living crocodylomorphs. The term (and its spelling variant 'crocodilian') has a complicated history and has at times been used for all crocodylomorphs.

Crocodylomorphs The archosaur clade that includes crocodylians and numerous fossil groups (including thalattosuchians) that are more closely related to crocodylians than to other archosaurs.

Diapsids The enormous reptile clade that includes lepidosaurs and archosauromorphs as well as numerous extinct groups related to these two. Diapsid means 'two openings' and refers to the two skull openings present behind the eye socket in these animals.

Dorsal fin A vertical fin, usually triangular or rounded, that grows from the midline of the back in many aquatic animals. Many fish have two dorsal fins, but aquatic reptiles only ever have one. The fin's function is probably to help keep the animal stable as it swims.

Estuarine Referring to objects, conditions and habitats that occur within estuaries: the places where large rivers meet the sea, and where fresh and salt water mix.

Evolution The process in which organisms change across the generations, the changes being heritable and passed from parent to offspring, and with the persistence of the changes varying according to the process of natural selection.

Fenestra In anatomy, a large opening in the skeleton, surrounded by bone. The plural term is fenestrae.

Flipper The wing-shaped limb of an aquatic reptile or mammal, used in generating lift as well as thrust (as opposed to a paddle, which only generates thrust). Flippers have bones and muscles along their length and are different from fins, which lack this sort of internal support.

Gondwana The great southern supercontinent that existed during the Jurassic and Cretaceous. During the Cretaceous, Gondwana split into Antarctica, Australasia, Africa, India, Madagascar and South America. The alternative name Gondwanaland is sometimes used.

Hupehsuchians An unusual Triassic marine reptile clade, so far known only from China. Hupehsuchians were mostly around 1 m (3¼ ft) long, had paddle-like limbs, toothless jaws, a stiff body, and a laterally compressed tail.

Hypothesis An idea put forward as an explanation for an observation, with there being some data that might allow other people to test its success as that explanation.

Ichthyosaurs A Mesozoic marine reptile clade that evolved a shark-like shape but began their history as thalattosaur- or hupehsuchian-like animals. They ranged from 1 m (3¼ ft) to over 20 m (65½ ft) and the majority had long, slender jaws and conical teeth. Ichthyosaurs originated in the Triassic and became extinct mid-way through the Cretaceous.

Ichthyopterygians A term originally used as the formal name for the ichthyosaur clade. More recently, some experts have used a naming system where Ichthyosauria is a clade within Ichthyopterygia, and where Triassic animals like *Grippia* and *Chaohusaurus* are within Ichthyopterygia but outside Ichthyosauria. However, these same experts still refer to all members of Ichthyopterygia as 'ichthyosaurs'.

Jurassic The section of the Mesozoic that lasted between 201 and 145 million years ago. Dinosaurs large and small dominated Jurassic life. Climates were seasonal but mostly tropical. Pangaea split into distinct northern and southern continents termed Laurasia and Gondwana.

Juvenile An organism that has not reached the full adult condition for its species.

Laurasia The great northern supercontinent that existed during the Jurassic and Cretaceous. It was separated from Gondwana by the Tethys Ocean. During the Cretaceous, the formation of the Atlantic caused Laurasia to split into North America and Eurasia.

Lepidosaurs The diapsid reptile clade that includes rhynchocephalians and squamates. Lepidosaurs originated during the Permian from an ancestor that also gave rise to archosauromorphs.

Lineage A branch of a family tree. A lineage can be enormous and include millions of species, or it could be short and include just a set of individuals within a single species. It effectively has the same meaning as 'clade'.

Mesosaurs A small Permian reptile clade – not to be confused with mosasaurs – of controversial evolutionary position. Mesosaurs ranged from 60 cm to 2 m (24 in to 6½ ft), and were long-jawed, long-tailed and probably amphibious.

Mesozoic The great section of time – the so-called 'Age of Reptiles' or 'age of middle life' – that lasted between 252 and 66 million years ago. The Mesozoic is divided into three subdivisions: the Triassic, Jurassic and Cretaceous.

Metacarpals The typically long, cylindrical bones that form the palm section of the hand and are located between the carpals (the wrist bones) and digits. Over the course of evolution, the metacarpals of several Mesozoic marine reptile groups became block-like in shape.

Metatarsals The typically cylindrical bones that form the middle section of the foot (the sole in humans) and are located between the tarsals (the ankle bones) and digits. Over the course of evolution, the metatarsals of several Mesozoic marine reptile groups became simpler in shape.

Mosasaurs A Cretaceous amphibious and aquatic lizard clade – not to be confused with mesosaurs – that originated as coastal, amphibious predators less than 1 m (3¼ ft) long but evolved into aquatic giants, some exceeding 15 m (49 ft). The most familiar mosasaurs had flippers, a vertical tail fin, and jaws and teeth suiting for the disabling of large prey.

Nothosaurs A Triassic sauropterygian clade, mostly limited to Europe and China. They often had long, shallow skulls with interlocking fangs, a flexible neck, and flipper-like limbs. They range from 1 m (3¼ ft) to 6 m (19¾ ft).

Pachypleurosaurs A Triassic sauropterygian clade, mostly limited to Europe and China. They are mostly less than 1.2 m (4 ft) long and have a flexible neck, pointed teeth and limbs suggesting amphibious habits.

Palaeontology The science of studying life of the past, practised by scientists termed palaeontologists. Palaeontology encompasses the study of ancient microfossils, plants, animals, traces left by organisms, and ancient environments and communities.

Pangaea The ancient supercontinent present during the Late Palaeozoic and Triassic. It broke apart into northern and southern sections (termed Laurasia and Gondwana) during the Jurassic.

Panthalassa The vast ocean that surrounded Pangaea during the Palaeozoic and part of the Mesozoic and, at times, covered around 70% of the Earth's surface. Plate movement led to the creation of new seafloor in the east of Panthalassa, and it is this expanding 'eastern' area that formed the Pacific Ocean.

Parvipelvians The shark-shaped ichthyosaur clade that includes *Ichthyosaurus* and its similarly-shaped relatives of the Jurassic and Cretaceous. Parvipelvians have small hind limbs, a small pelvis, and a shorter tail than other ichthyosaurs.

Phalanges The typically rectangular or cylindrical bones that form the digits (fingers and toes). Reptiles typically possess three or four phalanges in each digit but Mesozoic marine reptiles tended to increase this number over the course of their evolution.

Phylogeny The history of a given organism's evolution. The term phylogeny is also used for the diagrammatic trees we use when depicting a given hypothesis about evolution.

Physiology Everything pertaining to the way an organism functions – how it regulates and maintains its internal workings, including temperature control, water balance and salt balance, how energy is used, how it grows, and so on. The term is used both for the biological processes themselves, and for the science devoted to their study.

Pistosaurs A Triassic sauropterygian group, mostly limited to Europe and China. They were 2 to 3 m (6½ to 9¾ ft) long. Their long necks, flipper-like limbs and short tails reveal a close affinity with plesiosaurs, and it currently seems that plesiosaurs evolved from among pistosaurs. For that reason, pistosaurs as currently recognised are not a clade.

Placodonts A Triassic sauropterygian clade, mostly limited to Europe and China. They possess flattened or domed teeth on the jaw edges and palate and appear to have been predators of shellfish. Many species had armour plates covering the body and hence had a turtle-like appearance.

Plesiosauromorphs The informal term used for those plesiosaurs with a long neck and relatively small skull. This was probably the ancestral shape for plesiosaurs, and it was retained by many plesiosaur clades.

Plesiosaurs A major sauropterygian clade notable for their two pairs of flippers and plate-like limb girdles. They include long-necked forms (termed plesiosauromorphs) and short-necked forms (termed pliosauromorphs), as well as species intermediate between these extremes. Some plesiosaurs inhabited brackish and freshwater environments. Plesiosaurs originated in the Triassic and persisted to the end of the Cretaceous.

Pliosauromorphs The informal term used for those plesiosaurs with a short neck and relatively large skull. Pliosauromorphs evolved several times from different plesiosauromorph ancestors.

Pterosaurs An extinct archosaur clade, present throughout the Mesozoic, that contains the famous membranous-winged reptiles often called 'pterodactyls'. Pterosaurs were close relatives of dinosaurs within the archosaur clade Ornithodira.

Reptiles The major vertebrate clade that includes turtles, lizards, snakes, archosaurs and all of their relatives. The scientific meaning of the term is somewhat different from common usage, since 'reptiles' in the scientific sense includes birds.

Rhynchocephalians The lepidosaur clade that includes the Tuatara of New Zealand as well as numerous fossil relatives, most of which are Mesozoic in age. Rhynchocephalians originated in the Triassic from an ancestor that also gave rise to squamates.

Sauropterygians A Mesozoic marine reptile clade that originated and diversified during the Triassic, giving rise to placodonts, nothosaurs and plesiosaurs. Only plesiosaurs survived past the Triassic and into the Jurassic and Cretaceous.

Species A population of organisms where all individuals share features not present in other populations, and which generally all look alike and are all capable of breeding with one another.

Squamates The lepidosaur clade that includes all lizards and snakes (technically, snakes are a modified group of lizards). Squamates originated in the Triassic from an ancestor that also gave rise to rhynchocephalians.

Superclade An informal term for an especially big clade that contains several large clades. The term is redundant (since the vast majority of clades contain smaller clades) but has been used in the context of Mesozoic marine reptile studies to emphasise the size of the clade concerned.

Tail fin An expanded vertical structure (also called a caudal fin) that increases the surface area of the tail and assists in swimming. The term applies to long, continuous structures present along the length of the tail as well as to tall, triangular structures limited to the tail's terminal portion, to intermediates between these two, and to square, rounded and forked structures at the tail's tip too.

Tethys Ocean An ocean that existed between the Triassic and the Oligocene (between 250 and 20 million years ago) between the south-eastern margin of Laurasia and north-eastern margin of Gondwana. As these landmasses diverged, the Tethys Ocean spread westwards, forming the Western Tethys Ocean over Europe and northern Africa, and the Eastern Tethys Ocean over southern Asia. The Mediterranean, Black and Caspian seas are remnants of the Western Tethys Ocean. The Eastern Tethys Ocean closed as India moved north, colliding with mainland Asia.

Thalattosaurs A Triassic marine reptile clade – not to be confused with thalattosuchians – that includes long-jawed and short-jawed species. They ranged from 1 m (3¼ ft) to almost 5 m (16½ ft) and are long-tailed with limbs suggesting an amphibious lifestyle.

Thalattosuchians A mostly marine Mesozoic crocodylomorph clade that includes superficially gharial-like species as well as aquatic species equipped with flippers and a vertical tail fin. Thalattosuchians are sometimes called 'sea crocodiles' but they are not crocodiles nor closely related to them. The informal term 'sea crocs' is more appropriate.

Triassic The section of the Mesozoic that lasted between 252 and 201 million years ago. Thalattosaurs, ichthyosaurs and sauropterygians originated during the Triassic. The world was hot and huge deserts covered much of Pangaea, the lone supercontinent of the time.

FURTHER INFORMATION

Benton, Michael J. *The Reign of the Reptiles*. Kingfisher Books, 1990.

Ellis, Richard. *Sea Dragons: Predators of the Prehistoric Oceans*. University Press of Kansas, 2003.

Everhart, Michael J. *Oceans of Kansas: A Natural History of the Western Interior Sea (Second Edition)*. Indiana University Press, 2017.

Howe, Steve R., Sharpe, Tom and Torrens, Hugh S. *Ichthyosaurs: A History of Fossil 'Sea-Dragons'*. National Museum of Wales, 1981.

McGowan, Christopher. *Dinosaurs, Spitfires and Sea Dragons*. Harvard University Press, Cambridge, Mass., 1992.

Smith, Adam S. and Emmett, Jonathan. *The Plesiosaur's Neck*. UCLan Publishing, 2021.

Sues, Hans-Dieter. *The Rise of Reptiles: 320 Million Years of Evolution*. Johns Hopkins University Press, 2019.

Websites

Oceans of Kansas Paleontology http://oceansofkansas.com/

The Plesiosaur Directory https://plesiosauria.com/

INDEX

Page numbers in *italic* refer to illustration captions; those in **bold** refer to main subjects of feature pages.

Abyssosaurus 142, *142*
Acamptonectes 121
Acostasaurus 137
acromion 118
aeolodontines 157
Aequorlitornithes 81
Agassiz, Louis 63
aggression 112, 167, 173
aigialosaurs 165, 166, 169, 170
Aigialosaurus 170
albatrosses 80
Albertonectes 144, *144*
Alexander, Annie 84
Alexandronectes 145
Alienochelys 181
alligator lizards 167
Allopleuron 183
ammonites *105*, 151, 173, 177, 180
Amotosaurus 78
amphibious capability 98
anatomy 36–53, 127–8
Ancient Marine Reptiles 27
Andrews, Charles W. 147, 148
Andrianavoay 157
Angola 13, 26, 176, 177
Angolasaurus 176
Anguilla 37
Anguimorpha 167
anguimorphs 167, 169
Ankylosphenodon 90–1
Anning family 18, 20, 110, *113*
Anning, Mary **18–19**, *20*, 132
Anningsaura 19
Anshunsaurus 57
Antarctica 13, 16, 25, 145, 147, 168
antorbital fenestra *79*
apex predators 35, 73, 112, 122, 134, 143, 150, 151, 157, 162–3, 164
Aphanizocnemus 169
Archaeonectrus 132
Archelon 179–80, *180*, *181*
Archosauria 30
Archosauromorpha 30, *31*, 54
archosauromorphs 76
archosaurs 32
Arctic 108, 133
Arctic Ocean *see* Boreal Ocean
Argentina 12, 26, 118, 120, 121, 133, 150, 160, 162, *162*
Aristonectes 145, 146, 147
aristonectines 40, 47, *145*, 145–7, *146*
armoured reptiles 8, 36, 67, 68, 86, 87, 178
Arthropterygius 120
artistic reconstructions *18*, *46*, 52–3, *53*, *98*, *119*, *124*, *131*, *142*–3, *145*, *159*, *172*
Askeptosaurus 33, *84*, 85
Atacama Desert 113
Athabascasaurus 121

Atlantic 13, 15, 138, 185
Atopodentatus 64, *64*–5, *65*
Attenborosaurus 136
Attenborough, Sir David 136
Atychodracon 44
Augustasaurus 75
aulacodonty 38
Auroroborealia 108
Australia 25, 121, 137, 138, 148, 149, 150, 180
Austria 72, *79*, 80, 105

Bakker, Robert 27, 126
balance, sense of 45
Baptanodon 120
Baracromia 118
basking 155, 160
Baur, Georg 101
Belgium 21, 114, 116, 155, 173, 174, 177, 183
Benson, Roger 126
Besanosaurus 108
Birch, Thomas *16*, 17
birthing 100, *101*, *150*, 161, *168*
birthing sites 118
Bishanopliosaurus 133
bite 43, 73, 112, 115, 122, 135, 177
Bobosaurus 75, 76
body shapes *37*
 crocodile-like 152–63
 lizard-like *29*, 73, 164–77
 long-necked, big mouthed 124–51
 shark-like 4, 6, 94–123
 turtle-like 178–85
body size
 up to 1m (3¼ft) 5, 9, *16*, 57, 60, 72, 76, 82, 99, *100*, 164, 169, 179, 183
 1–5m (3¼–16½ft) 6, 8, 9, 57, 63, 67, 70, 73, 76, *77*, 78, 80, 85, 87, 88, 94, 95, 99, 101, 102, 103, *105*, 118, 119, 123, *136*, 138, 140, 149, *152*, 162, 171, 174, 177, 183, 185
 6–10m (19¾–33ft) *5*, 77, 94, 106, *106* 110, 112, 119, 127, 132, 134, 137, 138, 147, 157, 162, *163*, 175
 11–15m (36–49¼ft) 95, 106, 137, 142, *144*, 147, 172, 176, 177
 16–20m (52½–65½ft) 5, 103, 164, 176
 21–30m (69–98½ft) 105
 body temperature regulation 15, *72*, 155, 159, **160**
Bonarelli, Guido 35
bone density 36
Boreal Ocean 12, *12*, 14, 16, 103, 120, 133, 134, 160
Borealonectes 133
Bositra 23
Bosnia-Herzegovina 91
brachaucheniines 137–8
Brachauchenius 137
brachypterygiids *see* platypterygiines

Brancasaurus 148
Brazil 13, 58
Brazilosaurus 57
breaching 15
breastbone 49
British Museum 21
Buckland, William 124
Bunker, C.D. 173
Bunker mosasaur 173
buoyancy 36, 49, 131

Cabot, Godfrey L. 137
Cabrera, Ángel 145, 146
Caldwell, Michael 168
Californosaurus 108
Callaway, Jack 27
Callawayia 108
camouflage 93
Camper, Adriaan 16
Canada 77, 99, 102, 105, 116, 121, 133, 143, 144, 175, 177
 Alberta 149
 British Columbia 12, 73, 84, 85, 106, 107, *107*, 108
cannibalism 173
Caranx 37
carapace 67, 69, 178, 179, 183, 184, 185
Cardiocorax 144
Carinodens 177
Carroll, Robert 27, 62, 83
Carsosaurus 168, *168*
cartilage 48, 50
Cartorhynchus 31, 98, *98*, *99*
catfishes 166
Chacaicosaurus 118
chaohusaurs 99, 100, 101
Chaohusaurus 100, *100*, *101*, 103, 105
Charitomenosuchus 157
chelonids 183
Chelonioidea 178
chest bones 47
chevrons 46
Chile 26, 113, 145, 162
chin 139
China 11, 31, 33, 68, 70, 73, 74, 75, 76, 77, 78, 79, 80, 81, *85*, 86, 94, 98, 99, 102, *102*, 105, 106, 107, 108, 133, 155
 Guizhou 26, *54*, 57, *77*, 87
 Yunnan 87
clams 177
Clarazia 85
claudiosaurs 54
Claudiosaurus 60, 60–1
Clidastes 177
 propython 177, *177*
coelacanths 144
cold tolerance 13, 15, 160
Colombia 26, 121, 137, 143, 158, 171, 173, 180
colouration and markings 53, *111*, *119*, *141*, *184*
communication 45
Concavospina 85
Coniasaurus 169
Contectopalatus 102, *102*

continental fragmentation 11–14, 57
Conybeare, William 17, 20, *20*, 21, 110, 129
Cope, Edward Drinker 21, 22, *22*, 172, 175, *177*, 179
Cope's rule 172
coracoid 47, 49, 62, 75, 76, 103, 144
Corsochelys 184
cranial kinesis 43, 176
Creisler, Ben 75
crests 43, 102, 139, *147*, 148
Cretaceous 4
 extinction events 35
 sea snakes 91–3
crinoids *54*, 82
Croatia 170
Crocodylomorpha 152
Crocodylus porosus 153
Cruickshank, Arthur 44, 135
cryptoclidids 139–42
Cryptoclidus 30, 139–40, *140*
Crystal Palace models 6
ctenochelyids 183
Cuba 139, 162
Cuvier, Georges 17, 21, 22
cyamodontoids 67–9
Cyamodus 67, *68*
cymbospondylids 103–5
Cymbospondylus 103
 petrinus 105
Czechia 162

Dakosaurus 161, 163
 andiniensis 162
 maximus 162
De le Beche, Henry *18*, 19, 20
decompression syndrome 181
defensive adaptations
 saw-backs 183
 scutes 83
 tail shedding 59–60
dermochelyids 184
Dermochelys 185, *185*
 coriacea 184
Diandongosuchus 80
diapsid lineages 29–31
Diapsida 28, *31*
diapsids, early 60–1
Diedrich, Cajus 66
diet 5, 40
 ichthyosaurs 41, 107, 108
 mosasaurs *175*, 177
 pistosaurs 76
 placodonts 66
 plesiosaurs 125, 135, 138
 sea turtles 181
 thalattosaurs 85
 see also herbivory; prey *and* stomach contents
digestion 131, 170
digits 50, 109, 111, 114, 128, 165, 166, 173
 increase in number 82, 95, 116–17, 119
 webbed 57, 82, *88*
 see also phalanges
Dinocephalosaurus 78, 78–9

discoveries, fossil *see* fossil discoveries
distribution
 global 103, 125, 164
 restricted 113, 185
 role of seaways 12–14
diversity
 elasmosaurids 143–4
 ichthyosaurs 95–7
 metriorhynchids 162–3
 mosasaurs 165–6
 diving 44, 109, 142, *142*, 181
Djupedalia 142
dolichosaurs 165, 166, 169
Dolicorhynchops 151
Dollo, Louis 21, 174
Dong, Zhi-Ming 83
Donne, B.J.M. *19*
dorsal fins 6, 38, 51, 94, 100, 102, 109
drinking 168
Druckenmiller, Patrick S. 126

ear anatomy 44–5, 70, *139*
Ectenosaurus 175
Edgarosaurus 150
eels 37, 166
egg-laying *see* oviparity
Egypt 26
Eiectus longmani 138
elasmosaurids 48, 126, 142–7
elasmosaurines 144
Elasmosaurus 142–3, 144
 platyurus 21, 22, *22*
Enaliosauria 31
Enchodus 143
Endennasaurus 85
endothermy 15, *72*, 160, 168, 184
England 68, 117, 118, 135, 162, 169, 179
 Dorset 17, 18, 19, 112, 116, 136, 155
 Gloucestershire 105
 Lincolnshire 16
 Norfolk *123*
 Somerset 116
 Yorkshire 16, 110, 112, 114, 132, 139, 155
 see also Kimmeridge Clay; Lias; Oxford Clay; Purbeck Limestone *and* Wealden
Eohupehsuchus 83
Eonatator 173
epipodials 127
Eremiasaurus 176
Eretmorhipis 82, 83, *83*
Ethiopia 157
Euelasmosaurida 144
Eupodophis 92
eurhinosaurs *see* leptonectids
Eurhinosaurus 114, *114*, 115
eurysternids 179
Eusaurosphargis 88
Everhart, Mike 131
evolution
 convergent 79, 83, 107, 154, 167
 repeated *37*, 79, 126, 127,

INDEX

142, 143, 166, *166*, 170
turnovers and extinctions 32–5
Excalibosaurus 114
extinction, concept of 17
extinction events
  Cretaceous-Paleogene (K-Pg) 35, 185
  End-Permian (Permian-Triassic) 32, 33, 95, 185
  Jurassic-Cretaceous boundary event 34–5, 117
  Late Cretaceous (Cenomanian-Turonian; C-T; Bonarelli event) 35
  Late Triassic (Triassic-Jurassic; Tr-J) 33, 76, 109
  Middle Triassic (Ladinian crisis) 33, 34, 76
extinctions 11, 81, 86, 97, 160, 163, 164, 176, 178
  role in evolutionary turnovers 32–5
eye sockets 99, 101, 102, 103, 105, 142, 159
eyeballs 175
eyes, large 135, 142, *142*, 159, *176*

family tree, placement in reptile 5, 28, *31*, *32*
  hupehsuchians 83–4
  mesosaurs 58–9
  mosasaurs 166–7
  saurosphargids 87
  shastasaurs 108
  thalattosuchians 153–4, *154*
  see also superclade hypothesis
family trees
  ichthyosaurs and kin 97, 112, 115, 121
  mosasaurs 52, *166*, 171, 172, 177
  nothosaurs 72
  plesiosaurs 126, *126*, *127*, 133, 139, 143
  sauropterygians 64–5, *65*
  sea turtles 183, 184
feeding behaviour
  ambush 110, *135*, 140, 151, *157*, 171
  bottom-feeding 66, 87, 143
  crushing 85, 89, 181
  filtering and sieving 40, 58, 69, 146, 147
  grabbing 58, 85, 162, 174, 183, 184
  open-mouth attack 137
  prying 85
  suction 71, 73, 98, 99, 107, 181
feeding and teeth 38–43
Fernández, Marta 160
Fischer, Valentin 97, 121, 151
flightlessness 81
flippers 50, 124, 127–8, *128*, *129*, 130, 142, *165*, 173, *177*, 181, 184

fontanelle *179*
fossil discoveries 16–27, *97*
fossil misinterpretations 21–2, *22*, 52–3, *53*, 64, 137, 157
France 21, 77, 89, 105, 109, 110, 112, 114, 118, 120, 134, 139, 155, 157, 162, 179; see also Muschelkalk
freshwater habitats 62, 80, 123, 133, 147, 148, 149, 155, 171, 179
Frey, Eberhard 130
frigatebirds 80, *81*
'frills' 52–3
Fuyuansaurus 79

gaping 160
Gasparini, Zulma 160
gastralia 47, 49, *49*, 66, 82, *83*, 86, 125, 127, *142*
gastroliths 131, *131*, 151, 170
Gavialimimus 176
geckos 167
Geological Society of London 20
geological time, Mesozoic 4, *11*
geosaurines 162–3
Geosaurus 21, 25, 162, 163
Germany 21, 27, 68, 69, 70, 77, 78, 86, 89, 101, 102, *102*, 105, 116, *118*, 120, 136, 143, 148, *153*, 155, 157, 159, 161, 162, 169
  Holzmaden 23–4, 26, 47, 51, 109, 110, 114, 118
  Solnhofen 24–5, 90, 179
  see also Muschelkalk
gharials *154*, 155
gila monsters 167, *167*, 168, 169
Globidens 177, *177*
Gondwana 11, 13, *13*, 77, 138
Goronyosaurus 176
Gracilineustes 153, *158*
Grendelius 123, *123*
Griebeler, Eva 74
Gripp, Karl 99
Grippia 99, *99*
Grippidia 99
grippidians 96, 99, 101
Guanlingsaurus 54, 106, *106*, 107
guild, feeding 40–1, 102, 107, 110, 135, 157
Guizhouichthyosaurus 107, *108*
gulls 80
Gunakadeit 86

Haasiophis 92
habitats
  cold seas 13, 15, *15*, *142*, 149, 160, 184
  deep waters 142
  estuaries 98, 123, 147, 148, 149, *149*, 179
  fjords 49
  lagoons 69, 80, 147, 148, 149
  lakes 80, *149*, 155, 171
  oceanic surface waters 103

reefs 71
rivers 80, 155, 171
seafloor 10, *83*, 87, 110, 143
shallow coasts and seas 10, 66, *74*, 80, 99, 138, 179
shores 78
twilight zone 120
halisaurines 173–4
Halisaurus 173
Halstead, Beverly 27
hammerheads 65, *65*
'Harvard mount' 138
Harvard Museum of Comparative Zoology, Massachusetts, USA 137
Hastanectes 150
hatchlings 182
Hauff, Bernard 23
Hauffiopteryx 118
Hauffiosaurus 136
hearing 44–5
helveticosaurs 54, 86, 87–8
Helveticosaurus 87–8, *88*
Henodus 69, 69
herbivory 90, *91*, 183
Hescheleria 85, *85*
hesperornithines *81*
heterodonty 122
Himalayasaurus 106
hindlimbs 91, 92
Hispanic Corridor 12
Hobetsu Museum, Mukawa 185
Home, Sir Everard *16*, 17, 110
Houston Museum of Natural Science 183
hovasaurids 62
Hudson Seaway 14
Hudsonelpidia 109
Huene, Friedrich von 58, 86
Hungary 68, 159, 171
Hunter, John 17
hupehsuchians 9, 31, 54, 81–4, 94
Hupehsuchus 82
hydrodynamically-driven underwater olfaction 44
hydropedal condition 165, 166, *166*, 168, 170, 171, 172, 174
hydropelvic condition 165, 166, 172
hyperphalangy 50, *75*, 116–17, 119, 128

Ichthyopterygia 96
Ichthyosauria 96
ichthyosaurids 115–17
Ichthyosauromorpha 31
ichthyosauromorphs 31
ichthyosaurs 6, 6, 12, *16*, 17, 18, 23, 25, *34*, *54*, 94–5
  anatomy 38, *46*, 47, 49, 50
  diet 41, 107, 108
  diversity and history 95–7
  evolution 29, 30, *30*, 32–3
Ichthyosaurus 17, 20, 24, 101, 105, 109, 111, *114*, 116
  anningae 19

breviceps 117
communis 116
platyodon see *Temnodontosaurus somersetensis* 117
Igdamanosaurus 177
iguanas 167, 169; see also marine iguanas
India 13, 77, 148, 155, 157
Indonesia 102
Iraq 117, *117*
Israel 68, 73, 77, 92, 170, 177
Italy 16, 23, 68, 72, 73, 75, 77, 78, 80, 85, 88, 102, 105, 162, 171

jacks 37
Jaekel, Otto 24
Japan 95, 99, 150, 151, 173, 180, 182, 185
jaw muscles 43, 135, 143, *145*
jaws 96
  lower 43, *79*, 132, 164, *173*, *175*
  slender *59*, 108, 114, *114*
  upper 66, 181, *181*
  wide gape 122, 137
Johnson, Michela 157
Jordan *165*, 177
Jurassic 4
  extinction events 34–5
juveniles 51, 60, 67, 77, 88, *106*, *140*, 148, 160, 168

Kaiwhekea 145
Kallimodon 89
Kazakhstan 184
Keichousaurus 71
Ketchum, Hilary 126
Kimmeridge Clay 22, 24, 123, 134, 140
Kimmerosaurus 140
kingfishes 37
Klein, Nicole 74
Knight, Charles 52, *53*, 142
König, Charles 17
Kronosaurus 137, 138, *138*
Kyhytysuka 122

Labrador Seaway 14
land, movement on see terrestrial capability
Largocephalosaurus 87
Lariosaurus 72, 74
Laurasia 12
leatherback sea turtles 178, *184*, 184–5
Lebanon 92, 169
Lee, Mike 168
Leeds, Alfred 24
Leeds, Charles 24
Leidy, Joseph 21
Lemmysuchus 157
length, body see body size
Lepidosauria 30, 31, 54
leptocleidids 126, 147–9
Leptocleidus 147–8, *149*
  capensis 147
Leptonectes 109, 115

moorei 114
solei 114
leptonectids *34*, 50, 114–15
Leptosaurus 89
Li Chun 86
Lias 27, 47, 110
Libonectes 145
Libya 26
lightening adaptations 158, 179
limbs 37–8, 50, *115*; see also digits and flippers
Lingham-Soliar, Theagarten 26, 175
Liopleurodon 41, 134, 135, *135*, 136, *136*, 150
Loch Ness Monster 124–5
Lomax, Dean 117
Luskhan 138
Luxembourg 114, 155
Lydekker, Richard 22, 110

Macgowania 109
machimosaurids 157–8
Machimosaurus 157
  rex 157, *157*
Macroplacus 66
Macroplata 132
macrostomatans 93
Madagascar 60, 155, 157
Magyarosuchus 159
Maisch, Michael *30*, 59, 109
Makhaira 138
Malawania 117, *117*
mandibular fenestra 160
Mantell, Gideon 25
Maresaurus 133
marine iguanas 60, 91
marine life, adoption of 5–6, 89
Marsh, Othniel Charles 21–2
Massare, Judy 27, 40, 41, *43*, 107, 110, 135, 157
maturity, signs of 102
Matzke, Andreas *30*
Mauriciosaurus 47, 151
Mazin, Jean-Michel 27
McGowan, Christopher 27, 123
Megacephalosaurus 137
Merriam, John C. 84, 105, 106
Mesodermochelys 185, *185*
mesosaurs 54, 57–60
Mesosaurus 57
Mesozoic
  climates and temperatures 14–16
  environmental conditions 9–11
  geological time 4, *11*
  oceans and seas 11–14
Mesozoic marine reptiles
  anatomy 36–53
  characteristics 4–5
  evolution 28–35
  fossil discoveries 16–27
  lesser-known groups 54–93
  main groups 4, 6–9, 94–185
  metriorhynchids 12, *38*, 52, 153, *153*, *158*, 158–9
  diversity 162–3
  reproduction **161**

salt glands **160**
temperature control **160**
Metriorhynchus 162
Mexico 12, 26, 35, 150, 158, 162, 170
Meyer, Hermann von 22, 75, 76
microcleidids 138, 139
Microcleidus 44, 139, *139*, 143
Miodentosaurus 54, 85
mixosaurs 101–2
Mixosaurus 51, 101, 102
monitor lizards 17, 167, *167*, 168
Monquirasaurus 137
Monte San Giorgio 23, *24*, 27, 67, 70, *70*, 85, 87, 103
Mooreville Chalk 168
Morocco 26, *26*, 150, 155, *171*, 173, 176, 177, 180, 181
Morturneria 145, *145*, *146*
Mosasauria 165
mosasaurids 165, 170
mosasaurines 176
Mosasauroidea 165
mosasauroids 165
mosasaurs 9, 9, 13, 14, 21, 25, *26*, 35, 169–77
anatomy 36, *38*, 49, 50
biology and behaviour 167–9
diversity 165–6
in lizard family tree 166–7
Mosasaurus 6, 9, 17, 21, *26*, 172, 176
Motani, Ryosuke 31, 120
Mowry Sea 14
Mozambique Corridor 12
Muraenosaurus 143
Muschelkalk 72, 73; (France) *74*; (Germany) 22–3, 63, 67, 75, 76; (Netherlands) 67
Muscutt, Luke 130
Muséum national d'Histoire naturelle, Paris 16
Mystriosuchus 79, 80, *80*

Nakonanectes 144
Nanchangosaurus 82
Nannopterygius 120
nasorostrans 94, 98–9
National Museum of Ireland, Dublin *133*
Natural History Museum, London 21, 25, *113*, 132
Natural History Museum, Los Angeles *150*
neck
flexibility *48*, 71, 143
long *48*, *76*, 77, 78, *78*, 124, 125, 127, 142, 144
plesiosaurs **48**, 137
short 83, 125, 135, 137, 144, 149
Nectosaurus 85, *85*
neosuchians 152
nests 182, *182*
Netherlands 16, 73, 86, 88, 175, 177, 183; *see also* Muschelkalk
neural spines 46, 47, 48, 75, 82, 101, 108, 148, 159

Neusticosaurus 71
New Siberian Islands 108
New Zealand 25, 89, *89*, 145, 150, 177, 180
Nichollsaura 149
Niger 26, 173, 176
Nigeria 26, 173
nostrils 43, 44, 79, 99, 122, 150, 159, *181*
nothosaurs 8, 22, 23, 54, 72–5
Nothosaurus 57, 72, 73, 74
nyctosaurids 81

occultonectians 150
oceanic anoxic events 34
Ocepechelon 180–1, *181*, *182*
O'Keefe, Robin 126
omphalosaurs 96
Omphalosaurus 96
One Million Years B.C. 180
ophthalmosaurids 35, 50
ophthalmosaurines 119–20, *121*, *122*
Ophthalmosaurus 6, 24, 120, *121*
Ophthalmothule 142
Opisuchus 159
osteoderms 82, 83, 86, *87*, 88, 89, 155, 157, 159
Otschevia 123, *123*
overbite 114, *114*
oviparity 74, 168, 182
ovoviparity 60
Owen, Richard 22, 25, 31, *46*, 47, 62, 110, 124, 134
Oxford Clay 22, 24, 27, 120, *134*, 136, *136*, 139, *140*, 143, 157

Pachycostasaurus 136
pachyophiids 91
Pachyophis 91–2
pachypleurosaurs 8, 23, 44, 54, 63, 70–1
Pachyrhachis 92, *92*, 93, *93*
Pacific 13, 162, 185
paddle-like limbs 8, 50, 59, 69, 78, 82, 86, *100*,153, 158, 159, 164, 179
paedomorphosis 51, 142, 172
Palaeopleurosaurus 90
palate 44, 66, 73, 102, 132, 181
Palatodonta 67
Paludidraco 73
Pangaea 11, *11*, 57, 101
Pannoniasaurus 171
Panthalassa 11, 12, 73, 86, 103, 113, 133
Parahenodus 69
Paralonectes 85
Paraplacodus 66, 67
Parareptilia 58
parasites 159
parental care 60, 151
Parvipelvia 94
parvipelvians 6, 12, 24, 34, 36, *38*, 50, 108–9
pectoral girdle 47, 49, *49*, 62, 63, 71, 74, 127, 132

Pelagosaurus 152
Peloneustes 134, 136, *136*
pelvic girdle 47, *49*, 127
peripheral bones 185
Petrolacosaurus 29
phalanges 50, 74, 108, 116–17, 119, 128, 173
Phalarodon 102
Phantomosaurus 103
Phosphorosaurus 173
phytosaurs 54, 79–80
pistosaurs 8, 22, 54, *57*, 75, 75–6
Pistosaurus 75
placodonts 8, 22, 23, 36, *43*, 45, 54, *54*, 63, 66–9
Placodus 33, 66, 67, 68
gigas 63
plastron 67, 178, 179, *179*, 180
Platecarpus 21, 175, *175*
platypterygiines 119, 120–3, *122*
Platypterygius 120–1
Pleiosaurus see Pliosaurus
plesiochelyids 179
plesiopedal condition 166, 168, 169, 171
plesiopelvic condition 166, *166*, 171
Plesiopleurodon 150
Plesiosauria 127
Plesiosauroidea 126
plesiosauroids 125, 126
plesiosauromorphs 126, *126*, 136
plesiosaurs 6, *6*, 8, 23, 24, 25, 132–51
anatomy 36, 43, 44, 45, 47, *49*, 50, 127–8
behaviour and biology 131
classification 125–7
diet 125, 135, 138
discovery 18, *19*, *20*, 20–1
evolution 16, 27, *30*, 40, 125–7, *127*
in the Mesozoic environment 12, 14, 34, *34*, 35
neck **48**, 137
swimming styles 52, *129*, 129–30, *130*
Plesiosaurus 19, *20*, 20–1, 126, 132, 133, 136, 138, 142
Plesiosuchus 162, *162*, 163, *163*
Plesiotylosaurus 176
Pleurosauria *91*
Pleurosaurus 89, 90
goldfussi 91
plioplatecarpines 172, 174–6
Plioplatecarpus 21, 174, 175, *176*
pliosaurids 15, 25, 35, *41*, 134–6
Pliosauroidea 126
pliosauroids 125, 126
pliosauromorphs 126
Pliosaurus 124, 126, 135, 136
brachydeirus 134
Plotosaurus 176
Pluridens 173–4, *174*

Poland 72, 77, 86, 102, 155, 162
polycotylids 21, 126, 149–51
polycotylines 150
Polycotylus 149, *150*, 151
Polyptychodon 138
Pontosaurus 170
Portugal 91, 116, 155, 179
Posidonia 23
Posidonia Shale 22, 23, 109, 118
prefrontals 159
prey 41, 58, 73, 76, 78, 81, 85, 102, 103, *105*, 120, 140, 143, 151, 155, 157, 163, 180, 184
Primitivus 169, 170
Prionochelys 183
Prognathodon 21, *165*
currii 177
prolacertiforms 76
propodials 127, 128
Proteosaurus 16, 17
Proteus 17
Protoichthyosaurus 117
protorosaurs 54
Protostega 179, 181
protostegids 178, 179–82
Psephoderma 67
pteranodontids 81, *81*
pterosaurs 18, 80–1, *81*
Purbeck Limestone 123

Qianichthyosaurus 54
Quenstedt, Friedrich August von 23

rays 67
reproduction *see* birthing; oviparity; ovoviparity; unborn young *and* viviparity
retroarticular process 43
Reynoso, Victor-Hugo 91
Rhacheosaurus 153
Rhaeticosaurus 136
rhomaleosaurids *34*, 126, 132–3
Rhomaleosaurus 125, 132, *132*, 133, 135
cramptoni 133
rhynchocephalians 24, 57, 88–91
ribs 46, 49, 57, 70, 73, 74, 82, *83*, 86, 98, 99, 125, 136, 179, *180*, 183, 184; *see also* gastralia
Riess, Jürgen 49, 130
river dolphins 148
Robinson, Jane A. 129, 139
Romania 68, 73, 77
Romeosaurus 171
Romer, Alfred 137
rostrum 80, 84, 157, 180, *182*
Russell, Dale 27, 172
Russellosaurina 172
Russellosaurus 171
Russia 25, 102, 108, 120, 121, 122, *122*, 138, 142, 143, 149, 155, 162, 169, 180

Sachicasaurus 137
sacrum 46
salt glands 43, 65, 122, **160**, *161*
salt tolerance 60
saltwater crocodiles 153
Sander, Martin 106, 107
sapheosaurs 89
Sapheosaurus 89
Saudi Arabia 77
Sauropterygia 54, 62
sauropterygians 29, 31, 37–8, 44, 49, 62–3, 66–76
family tree **64–5**, *65*
saurosphargids 9, 31, 54, 86–7, *87*
Saurosphargis 86–7
scapula 47, 49, 62, 96, 101, 103, 118, 132, 144
scent detection 44–5, *45*, *174*
scleral ossicles *139*
Sclerocormus 98
Sclerostropheus 78
sclerotic ring 103, 120, 159, 175
Scotland 124–5
scutes 83, *83*, 178
sea conditions
acidity and oxygen depletion 11, 34, 35
sea levels 10, 11, 14, 86
sea temperatures 10, *10*, 15
sea crocodiles (sea crocs) 152–63
sea jellies 184
sea levels 10, 11, 14, 86
sea snakes 43, 57, 91–3
sea turtles 9, 24, 36, 43, 178–82
hard-shelled 183–4
leatherback 184–5
seabirds 43, 80, 81, *81*
Seeley, Harry 132
Seeleyosaurus 47
Selmasaurus 176
sexual dimorphism 71
sharks 144
shastasaurs 105–8
Shastasaurus 105, 106–7, *107*, 108
Shonisaurus popularis 105–6
sikanniensis 107
sight *see* vision
Simolestes 136, 148
simoliophiids 91
Simoliophis 91
Simosaurus 72, 73, *74*
Sinocyamodus 68
Sinosaurosphargis 87, *87*
size *see* body size
skeletons
ichthyosaurs and kin *99*, *100*, *110*, 111, *116*
mesosaurs *59*
mosasaurs 9, 171, *173*
nothosaurs *74*
placodonts 66
plesiosaurs *20*, 20–1, *49*, *133*, *134*, *136*, *140*
rhynchocephalians *91*
sea turtles *180*, *183*, *185*

thalattosaurs *84*
thalattosuchians *153*, 155
skin 50, 51, *51*, *165*, 175
  scaly 90, 155, 164, 170
  smooth 136, 159
  see also colouration and markings
skinks 167
skull
  anatomy 43–5, *147*, *162*
  evolutionary modification 28–30
  flattened *68*, 69, 140, 146, 171
  size evolution *100*, *105*
  skull specimens
    ichthyosaurs and kin *30*, *99*, *112*, *113*, *114*, *116*, *123*, *123*
    mosasaurs *175*, *176*
    nothosaurs *73*
    phytosaurs *79*
    placodonts *68*
    plesiosaurs *30*, *132*, *136*, *139*
    sea turtles *180*, *181*, *182*
    thalattosuchians *153*
Slovakia 155, 162
Slovenia 170
slow-worms 167
snakes 91, 93, 167, 168
snout 43, 44
  broad 132, *132*, 151
  down-curved 85, *85*
  flattened 78
  glands within 161
  grooved 109, *110*
  long 57, 67, 75, 90, 102, 114, 117, 118, 132, 134, 136, 138, 149, 157, *173*
  narrow 67, 73, 84, *84*, 102, 137, 138, 149, 162, 175, 180
  prow 172
  short 60, 67, 71, 72, 87, *88*, 98, 107, 114, 117, 118, 136, 137, 142, 174, 176
  slender 76, 79, 80, *100*, 102, 114, 136, 151, 176
  tubular 159
social behaviour 60, *141*, 169
soft tissues 23, 25, 45, 47, 50, 51–2, *116*, 128, *128*, *153*, *165*, 169
Solnhofia 179
Sömmerring, Samuel von 25
South Africa 26, *58*, 148
Spain 68, 69, 73, 77, 114, 162, 177, 179
Sphargis 86
Sphenodon punctatus 30, 89
Spindler, Frederik 159
Spitrasaurus 142
Spitsbergen 24, 27, 86, 88, 96, 99, 102, 103, 120, 142
Squamata 9, 166
stenopterygiids 118
Stenopterygius 24, 51, 53, 105, 109, *118*, *119*
  quadriscissus 118
  uniter 118
Stenorhynchosaurus 137

Stereosternum 57, *58*
stomach contents 41, 58, 81, 93, 102, 107, 108, *108*, 112, 121, 135, 140, 143, 151, 164, 173, 177, 181; *see also* gastroliths
strandings 106, 166
Stratesaurus 20, 132
stromatolites *58*
Stukeley, William 16
sturgeons 83
styxosaurines 144
Styxosaurus 144, 145
Suevoleviathan 109–10, *110*, *111*, 115
suevoleviathanids 109–10
Sulcusuchus 150
Sundance Sea 12, 120
superclade hypothesis 31–2
Sveltonectes 122
Sweden 173, 175, 177
swimming
  acceleration 49, 110, 116, 135, *135*, 182, 184
  braking 49, 50, 116
  fast 37, 38, 48, 51, 69, 96, 109, 115, 116, 151, 181
  lift 119, 129
  long-distance 37, 109, 110, 165
  slow 66, *69*, 71, *74*, 82, 87, *92*, 99, 115
  speed bursts 101, *135*
  stability 36, 38, 102, 119, 140, 158
  steering 50, 119, 158
  thrust 4, *22*, 50, 129, 130, 158
  turning 49, 87, 116, 131, 181
  see also diving
swimming styles 37–8, *38*, 52, 57, 71, 76, 87, 90, 96, *129*, 129–30, *130*
Switzerland 23, 68, 72, 77, 78, 88, 102, 114, 116, 146, 155, 162, 179
swordfish 115

tail 37–8, *46*, 85
  flexibility 59
  large *111*
  lobed *165*, 175
  long 90, *91*, *100*, 170
  short *125*, 127, 142
  vertebrae 101, 109, *110*, 170
  tail fins *38*, *153*
    crescent-shaped 6, 47, 159
    diamond-shaped 47
    horizontal 47, *145*, 146
    vertical 8, *9*, 37, 51, 100, 111, *158*, 164
tail shedding 59–60
Taniwhasaurus 25
tanystropheids 9, 23
Tanystropheus 23, *57*, 76, 76–8
Tarlo, L.B. *see* Halstead, Beverly
Tatenectes 140, *141*
taxonomic revisions 72, 155, 177
Taylor, Michael A. 27, 131, 135

teeth
  acrodont 89, 90
  button-shaped 96
  classification 40–1, *43*
  conical 85, 105, 108, 120, 164
  count (high) 64–5, 175; (low) 175
  fang-like *30*, *88*, 145
  hooked 177
  interlocking *30*, 145
  keeled 110, 164, 176
  large 106, *122*, 139, 143, 162
  multicusped 78
  projecting *146*, 177
  recurved 92, 140, 162, *176*
  replacement 40
  ridged 150
  rounded *43*, 63, 99, *177*, *177*
  slender 175
  striated *41*, *102*, 105
  variation 86–7, 89
  wear 90, 162
teeth and feeding 38–43
teleosauroids 52, 153, *154*, 155–8
Teleosaurus 21, 155, 157
temnodontosaurs 50, 110–13
Temnodontosaurus 17, *95*, 109, *110*
  acutirostris 112
  azerguensis 112, *112*
  eurycephalus 112, *112*
  platyodon 112, *113*
  trigonodon 112
temperatures
  air 14, 32
  sea (actual) 10, *10*, 15; (differences) 14–15
  for body temperature *see* body temperature regulation
Terminonatator 144
terrestrial capability 8, 9, *64*, 71, 75, 76, *76*, 78, 85, 86, 88, 98, 155, 159, 169
territoriality 173
Testudines 178
Tethys Ocean 5, 8, 11, *13*, 68, 73, 77, 81, 86, 94, 103, 113, 114, 120, 133, 134, 138, 157, 169
tethysaurines 170–1
Tethysaurus 171, *171*
Thailand 99, 155
thalassemydids 179
Thalassochelydia 179
Thalassomedon 145
thalassophoneans 136, 137
Thalatoarchon 106
thalattosaurs 9, 23, 31, 54, *54*, 57, 84–6, 107, *108*
Thalattosaurus 84–5
Thalattosuchia 30
thalattosuchians 6, 8, 16, 21, 23, 24, 34, 35, 152–3, 155–63

anatomy 36, 43, 49, 52
  in the croc family tree 153–4
thecodonty 38
Thililua 151
throat bones 71
Thunnosauria 115
Thunnus 37
tongues, forked *167*, 168
toothlessness 82, 89, *106*, 115, 118, 120
toretocnemiids 108
Torvoneustes 163
Toxochelys 183, 183–4
tracks 74
'tragsystem' 159
transverse processes 170
Triassic 4
  extinction events 32–4
  parvipelvians 108–9
  sauropterygians 62–76
Trinacromerum 151
tuataras 30, 89, *89*
Tübingen University, Germany 136
tunas 37
Tunisia 157
Turgaiscapha 184
Turkey 68, 102
turtles 28; *see also* sea turtles
tylosaurines 172–3
Tylosaurus 21, 52, 53, *53*, 172, 173, *173*
  proriger 172

Ukraine 177
Umoonasaurus 149
unborn young 23, 60, *60*, *70*, 71, 71, 78, 100, *101*, 105, 108, 118, *118*, *150*, 151, 168
underwater flight 49, 129, 175
undorosaurids *see* platypterygiines
UNESCO World Heritage Sites 23
Uruguay *60*
USA 33, 96, 102, 120, 121, 124, 144, 149, 151, 169, 170, 177, 182, *183*, 184
  Alabama 168, *177*, 183
  Alaska 86
  California 84, 85, 105, 106, 176
  Gulf Coastal Plain 21
  Kansas 21, 137, 168, 172, *173*, 175, 179, 183
  Montana 150
  Nevada 103, *105*, 105–6
  New Jersey 21, 173
  Oregon 158, 159
  South Dakota 177, 179
  Texas 171
  Wyoming 12, *140*, 150
Utatsusaurus 95, 99

Vadasaurus 90
Vallecillosaurus 170
Venezuela 138
venom 89, 168–9
vertebrae 21, 48, 77, 101, 103, 109, *110*
  up to 20 vertebrae 78, 148
  21–30 vertebrae 48, 71, 78, 132, 135, 148
  31–40 vertebrae 76, 138, 144, 148
  41–50 vertebrae 142, 143
  51–100 vertebrae *106*, 141, 144, 170
  101–150 vertebrae 21, *92*
vertebral column 46–7
  flexibility 173
  stiffened *121*
Viking Corridor 12, *12*
vision 135, *141*
  binocular 173
  field of vision 159
  low-light 120
viviparity 49, 60, 65, 71, 74, 78, 100, 160, 168
volcanicity 11, 84
vortices 130, *130*

Wahlisaurus 114
Wales 117
Walking With Dinosaurs 134
Wangosaurus 75, 76
'warm-bloodedness' 15, *72*, 160, 168, 184
Water Reptiles of the Past and Present 27
Watson, David M.S. 125, 129
Wealden 123, 147, 150
Weddellian Province 13
Wegener, Alfred 57
Welles, Samuel 125
Western Interior Sea 14, *14*, 142, 183
whales 107, 147, 166
White, Theodore 125
Williston, Samuel 27, 149
Wiman, Carl 99
Wimanius 102
windpipe 175
Wumengosaurus 31, 32
Wunyelfia 145

Xenodens 177
Xenopsaria 126–7, 149
Xinminosaurus 103
Xinpusaurus 85, *108*

Yaguarasaurus 171
Young, Mark 162
Yucatán Peninsula 35
Yunguisaurus 75
Yuzhoupliosaurus 133

Zammit, Maria 48
Zarafasaura 145
Zealandia 13
Zoneait 159
zygapophyses 48